T0073729

IMPERFECTION

IMPERFECTION

A NATURAL HISTORY

TELMO PIEVANI

TRANSLATED BY MICHAEL GERARD KENYON
FOREWORD BY IAN TATTERSALL

The MIT Press
Cambridge, Massachusetts
London, England

The MIT Press would like to thank the anonymous peer reviewers who provided comments on drafts of this book. The generous work of academic experts is essential for establishing the authority and quality of our publications. We acknowledge with gratitude the contributions of these otherwise uncredited readers.

This book was set in Adobe Garamond Pro by Jen Jackowitz. Printed and bound in the United States of America.

Library of Congress Cataloging-in-Publication Data

Names: Pievani, Telmo, author. | Kenyon, Michael Gerard, translator.
Title: Imperfection : a natural history / Telmo Pievani ; translated by Michael Gerard Kenyon ; foreword by Ian Tattersall.
Other titles: Imperfezione. English
Description: Cambridge, Massachusetts : The MIT Press, [2022] | Originally published: Imperfezione: Una storia naturale. Milano : Raffaello Cortina editore, 2019. | Includes bibliographical references and index.
Identifiers: LCCN 2021060546 (print) | LCCN 2021060547 (ebook) | ISBN 9780262047418 (print) | ISBN 9780262371650 (epub) | ISBN 9780262371667 (pdf)
Subjects: LCSH: Imperfection. | Evolution (Biology)—Philosophy.
Classification: LCC QH360.5 .P54 2022 (print) | LCC QH360.5 (ebook) | DDC 576.801—dc23/eng/20220412
LC record available at https://lccn.loc.gov/2021060546
LC ebook record available at https://lccn.loc.gov/2021060547

10 9 8 7 6 5 4 3 2 1

To Carlo, Gianluigi, Roberto, and Sandro,
the band that transforms imperfection into art

CONTENTS

FOREWORD

IAN TATTERSALL

Telmo Pievani has a healthy respect for the virtues of imperfection—a quality ordinarily more disdained than admired. That unusual respect derives not only from Pievani's perception that the world's imperfections are what make it an interesting place to live in, but also from his acute awareness that, in the absence of imperfection, the biosphere of which we are a part could never have evolved. For, as he elegantly shows, the history of the world—and indeed that of the universe of which it is an infinitesimal part—is in its essence a history of imperfection. In fact, he goes so far as to declare that "perfection [itself] is paradoxical." At a time when our ideas of ourselves and the world around us are so strongly conditioned by the (implicitly perfecting) technological "progress" that pervades our social existences, this unconventional view is a hugely salutary one—especially when it comes to investigating and understanding the origins and the nature of our own bizarre species, *Homo sapiens*.

At the beginning of his compact yet amazingly wide-ranging account of how the universe, biosphere, and human beings came to be as they are, Pievani underlines the evolutionary importance of historical contingency, whereby each significant new development in evolution opens up a finite range of new possibilities while eliminating others. Once you are on a new track, old options are closed to you, while your future opportunities will almost certainly be constrained not only by what history has arbitrarily handed you, but also by the myriad external influences that are entirely random with respect to the kind of creature you are. It is this inescapable reality that makes every evolutionary trajectory unique and almost certainly unrepeatable—even in principle. And while Pievani does note the sheer a priori improbability that, even after almost four billion years of life on earth, an organism should have emerged that walks on two legs, writes symphonies, and sends rockets to the moon, he wisely warns against the seductive hindsight that makes that emergence somehow seem inevitable, or part of a progression toward a perfection of some sort. After all, once the process was underway, an outcome of one kind or another was assured, and we are astonished by the one that eventuated simply because we are a creature with precisely those qualities, and because we happen to be capable of such astonishment.

Pievani cleverly constructs his book as a roughly chronological series of accounts of major evolutionary events, interspersed with discussions of the factors—ranging from genomic mechanisms to ecological constraints—that might most plausibly explain them. The volume you are holding is consequently as much a rumination about evolution itself as a chronicle of what happened as the biota changed over vast spans of time, ultimately giving rise to a biped that is capable—occasionally—of rational thinking. Perhaps most important, we learn that in evolution the important thing is not to be optimized for anything—after all, what is optimal is entirely relative in an environment that has historically been highly susceptible to change—but rather simply to be good enough to get by in whatever circumstances happen to present themselves. Different circumstances, different results; and then, there is always that historical contingency.

We see this nowhere more clearly than in the evolution of our own human lineage, which was diverse even in the quite recent past but boasts only one living species today. Pievani pays particular attention to the human brain, which greatly enlarged within various lineages of our genus *Homo* over the past two million years. In the end, though, it wasn't size that mattered most: at least some of the hominin relatives that *Homo sapiens* recently put out of business had brains just as large. What made the difference, and what explains our lonely state today as the world's only hominin, was how the brain functioned, especially in regard to reasoning and planning. But for all its effectiveness in the competitive arena, the modern human brain is structurally as well as functionally a rather messy instrument: neuroscientist Gary Marcus called it a "kluge"—an untidy device, jury-rigged from whatever was available, that has nonetheless turned out to do the job well enough. Clearly, it was the extraordinary properties of its brain that allowed *Homo sapiens* to eliminate all the hominin competition in a remarkably short time; but given our faulty memories, propensity to make bad decisions, and tendency to believe crazy things, it would equally clearly be hopeless to claim that evolution has optimized our brains for anything whatever. As Pievani trenchantly observes when he ponders why humans were able to take over the world, "Simply put, our imperfection functioned better than theirs."

This volume is, then, no run-of-the-mill survey of the history of the earth's biota and the human species. From the pen of one of Italy's most thoughtful and influential philosophers of science, it is a probing analysis, couched in spare and compelling prose, of just how a world of single-celled organisms came to transform itself, over eons, into the incredibly profuse and diverse biota familiar today—and of how one odd primate lineage came to dominate that biota. Of course, as its author would be the first to admit, any investigation of this kind must necessarily be a progress report rather than a definitive declaration. Yet, as the highly readable product of a profound but lightly worn erudition, this book will at the very least set you to thinking—and it might even change your view of your place in the world.

1

A SUBTLE IMPERFECTION: AND SO IT ALL BEGAN

> It is demonstrable that things cannot be otherwise than as they are; for all being created for an end, all is necessarily for the best end. Observe, that the nose has been formed to bear spectacles—thus we have spectacles. Legs are visibly designed for stockings—and we have stockings. Stones were made to be hewn, and to construct castles—therefore my lord has a magnificent castle; for the greatest baron in the province ought to be the best lodged. Pigs were made to be eaten—therefore we eat pork all the year round. Consequently they who assert that all is well have said a foolish thing, they should have said all is for the best.
> —Voltaire, *Candide, or Optimism*

In the beginning, there was imperfection. A rebellion against the established order, with no witnesses, in the heart of the darkest of nights. Something in the symmetry broke down 13.82 billion years ago. An imperceptible breeze started blowing, and disastrously the great pencil of the universe fell on one side and not on the other. And so an infinitely tiny anomaly became the source of all things.

A VOID, CONTAINING EVERYTHING

The extraordinary physical research carried out in recent years on the infinitely large and infinitely small, the product of bold predictions made in the last century, has converged toward the hypothesis that our universe is nothing more than the relentless metamorphosis of a perfect void. Yes, a void. A total absence of matter, fields, particles. Nevertheless, the void from which everything began was not pure nothing. On the contrary, it was everything. And this everything persisted in a state of energetic equilibrium. But this primordial void was not motionless; its energy fluctuated. It was a quantum vacuum, teeming with random oscillations, symmetrical collisions, and mutual annihilations between particles and antiparticles. Perfect in its overall energy equilibrium, but restless and bubbling. It contained everything, and the opposite of everything. This trembling void was the primordial matrix of all possible outcomes and stories. Although complete in itself, and therefore perfect, it was unstable.

Then something happened that the great Latin poet Lucretius, following his master, Epicurus, named *clinamen* in his *De rerum natura*. The first Greek atomists pictured the primordial state of the world as an eternal rain of particles that had in all eternity fallen regularly side by side and in parallel. But then this harmony snapped, and a tiny disturbance diverted the trajectory of one atom, which then hit another, and then another and another, in a chain reaction that disrupted the initial deterministic picture, triggering the history of the cosmos in all of its magnificent imperfection. And so it all began, through a small accidental deviation or derailment. Something similar could have happened, just as accidentally, in one of the infinite fluctuations taking place in the original quantum vacuum. It was, like many others, just a tiny fluctuation. Possibly in the presence of an elusive cousin particle of the Higgs boson, the inflaton, primeval symmetry broke down. This time, however, the perfect state of the vacuum was not restored. The equilibrium shattered, and an inflaton bubble, driven frantically by the energy of the void, exploded space-time at an incalculable speed.

That primal imperfection, caused by the inflaton rebellion, directed us toward the history of everything we know—if "history" is a term we can use, considering that it all took place in a few billionths of a second. After another of these imperceptible moments, the inflatons fed off themselves and the explosion expanded exponentially, generating a macroscopic, incandescent space-time filled with massless, lightning-fast particles that obeyed a single, unified force. Then this inflation suddenly slowed down, almost like an afterthought. For a trillionth of a second, the universal symmetry seemed to restore a state of apparent perfection; but this state of affairs ended almost before it started. After the paroxysmal inflationary phase, the gravitational force began to act, the drop in temperature caused the Higgs bosons to condense, the electromagnetic force split from the weak nuclear force, and the two in turn separated from the gravitational force. Particles started to interact with the ubiquitous Higgs scalar field, encountering different resistances and thus diversifying. These particles, having been slowed down by the field, acquired distinct masses: quarks, leptons, and force-mediating bosons.

And so another anomaly had imprinted itself on the world, like an irreversible signature. The result was an exotic biodiversity of elementary particles, including those that have survived to this day and those that have died out. This second asymmetry led to visible matter, or perhaps even to dark matter, light, and the four fundamental forces. In other words, the structure of the universe as we know it took shape. And only one immaculate second had passed.

ANISOTROPY

Simultaneously, a third, slight but fundamental imperfection influenced the course of events for all time. For reasons yet to be discovered but that might also be linked to the properties of the Higgs field, matter just managed to prevail over its specular counterpart, antimatter, resulting in the arbitrary excess of the former over the latter that we can still observe today. This infinitely small disproportion gave rise to the material, and not "antimaterial," nature of our reality.

What then followed was a rapid cascade of other asymmetries, ramifications, and aggregations: a plasma of quarks and gluons, then protons, electrons, neutrons, and the first charged nuclei, followed by atoms and molecules, hydrogen and helium. That split second led to a period of 380,000 years from the *beginning*—if "beginning" can be considered the correct term when there was no *before*—when light was finally able to separate from matter, and in a flash photons started to propagate freely everywhere. The cosmic microwave background radiation, which we still register today, is the fossilized signal of the first flash of our universe that became transparent. It has always permeated the cosmos, and reaches us from the sky in all directions. At first glance it seems homogeneous because it spread from the same point and in every direction, but this is not completely true. In fact, looking at its tiny ripples and at differences from various regions of the cosmos, we can perceive another important imperfection. The background radiation reveals a weblike structure.

When the hydrogen and helium clouds first came together, gravity did not act uniformly. Perhaps due to the presence of a cosmic network of dark matter, the first stars and galaxies formed in denser areas that were separated by immense, more rarefied, and empty areas. The miniscule initial inhomogeneities became centers of attraction around which the first star clusters and then galaxy clusters formed. This explains why the structure of the universe today is not uniform, and why not everything can happen everywhere. If we think about it for a moment, the underlying mechanism is Lucretius's clinamen, a break in symmetry. If matter had been uniformly distributed to perfection (i.e., was isotropic), gravity would have acted equally everywhere, and matter would have remained locked in its isotropy while space-time expanded. Should we, however, introduce a slight anomaly, a deviation, a perturbation of the isotropy, the gravitational attraction would be slightly stronger in one area than another. This anisotropy, although infinitesimal at the beginning, as in the clinamen, deepens and widens more and more because, as the matter becomes denser, it attracts other matter, the gravitational instability increases, and a diversified structure is created with deep dissymmetries from area to area. This is, in fact, what actually happened—all due to minimal imperfections that were perhaps already present at the time of the initial inflation, and were then amplified.

The anisotropy of the universe is another of its wonderful imperfections that is still visible today in the form of exceedingly small temperature differences (in the order of two hundred millionths of a degree) in the cosmic background radiation, which is a photograph of the universe at 380,000 years of age and a fingerprint of its nonhomogeneous structure. From that moment on, this configuration of asymmetries in matter density became a watershed that has conditioned all subsequent events. Like a ratchet in a gear, there is no turning back. In the parts of the universe with the highest density of matter, the expansion of space-time was slowed down by gravity. In the most unstable parts of the universe, gravity collapsed considerable masses of matter, the temperatures inside rose, and the first nuclear furnaces of protostars were ignited.

After 300 million years of darkness, the universe lit up with countless numbers of isolated flares. Without that anisotropic imperfection, carbon, nitrogen, oxygen, neon, sodium, magnesium, silicon, and gradually heavier elements such as sulfur, calcium, and iron would never have been synthesized in the heart of the stars (Baggott 2015). The collapse of those early stars, and the explosive energy released by supernovas, littered the interstellar clouds with even heavier elements. In this way, later, more stable and long-lived stars came into being with a more diverse chemical composition. In the random structure of the cosmos, superclusters of galaxies, great walls, majestic filaments, strings of stars, and nebulae were formed. Some areas contained heavy elements, and others less. And thus a great stage was set up hosting a drama of unpredictable outcomes. The history of the universe is a long sequence of asymmetries. What is needed is a theory of superasymmetry.

CONTINGENCY

The universe is also a dangerous place. Its vastness is shaken at every moment by violent catastrophes that would annihilate us instantly. It inspires a sense of sublime power, and yet the physics of the infinitely large and the infinitely small tells us that it, too, is precarious. For everything is precarious. The universe was born, has evolved, and will vanish, either through a slow, cold death or in a grand final explosion. We have known for decades that it has evolved, and that one day it will finally end. The fact that chance and instability have influenced its history is something we have been learning the hard way in recent years.

Everything is precarious because it is imperfect, unnecessary, and incomplete, and because it could have been different.

Lucretius's fruitful clinamen, revisited using the language of twenty-first-century science, becomes what we could call "turning points" or differences—sometimes tiny variations that have made a difference. These are historical conjunctures that have two characteristics: *they are unpredictable regarding what preceded them* because they are the result of complex nonlinear interactions between multiple interdependent factors that would have made other alternative scenarios possible; and *they are decisive for what happens afterward* in the sense that the bifurcations they introduce have a profound and causal influence on the course of subsequent events. In other words, they are unpredictable because the previous states of the system are necessary but not sufficient to predict them in advance; and, in turn, the future states of the system are causally dependent on them. The product of this process therefore can be seen as contingent in the sense that it depends heavily on its previous history, or, to put it another way, on the sequence, which is unique in each case, of the critical points that preceded it.

In this definition, most historical events are not critical moments either because they were unpredictable but had no significant consequences, or because they had a major impact on the future but were relatively predictable at the time they took place. If each of the countless episodes in an event were a critical moment, there would be no understandable history because pure and total chaos would dominate. It is thus normal that many episodes are irrelevant to the course of following events. On the contrary, however, if no episode in an event were a critical moment, it would be pointless to wait to see the outcome of history since everything would already be determined and written right from the beginning by the rigid laws of the invariant procedure.

The most wonderful stories that have been investigated by science fortunately fall into neither the first nor the second extreme case. The most interesting fall in between, and are stories in which laws and chance, as naturalist Charles Darwin said, interact from time to time in the most unexpected ways. In these situations, the course of events has its own logic and consistency—a logic dictated by regularities and underlying laws—but the robustness of the process (its obedience to recurring causes) is not so strong as to make it deterministic a priori and predictable. In other words, the process is punctuated by influential critical turning points or "frozen accidents" (contingent results of history that reverberate in future events) that alter its course, modify its outcomes, and transform it into something unique.

Contingency, then, refers to the causal power of a single individual historical detail, or, if you prefer, dependence on history. Sensitivity to the contingent event unites the most important processes that affect us directly: cosmic and biological evolution, individual development, and our own lives, with their decisive meetings and crucial turning points (Kampourakis 2018). In the incessant tug-of-war between laws and chance, contingency can manifest itself to different degrees. On some particularly fluid occasions, sensitivity is high and many critical points split open a range of alternatives—that is, equally probable paths for history to follow. On other, more crystallized occasions, sensitivity may be low because the invariant laws of the process or the constraints of the moment make certain outcomes much more likely than others. Which phase of your life are you in right now? Fluid or crystallized? High or low contingency?

LET'S KILL HINDSIGHT

Everything seems to indicate that the degree of contingency in the cosmic and planetary histories that concern us has always been quite high. Within the imperfect and therefore fascinating cosmic scene, our local neighborhood is in no way special. We are 27,000 light-years from the center of a normal spiral galaxy, the Milky Way, in the middle of one of its peripheral arms, the Orion spur. With its minimum 100 billion stars, our galaxy is part of a modest cluster of 50 galaxies known as the Local Group, which is itself 1 of 100 that make up the Virgo supercluster, and will collide with the Andromeda galaxy in approximately 400 million years' time. Our unimportant region is, however, 10 billion years old, which means it is old enough for many first-generation stars to have had time to exhaust themselves and explode into supernovas, with the consequential spread of a rich array of heavy elements—which is what matters to us. It is a favorable local contingency, the result of the previous sequence of critical moments.

In our local region, stardust, the soup of heavy element molecules, included a wide assortment of carbon compounds. The organic molecules circulating also included amino acids, nitrogenous bases, and other interesting aggregations in the form of linear chains of molecules. With the exception of hydrogen, all the matter that our bodies are made up of came from these interstellar chemical factories. About 4.8 billion years ago, around the time when space-time again began to accelerate its expansion due to the countergravitational effect of dark energy, a cold and dark interstellar cloud on the outlying regions of the Milky Way started to collapse. As philosopher Immanuel Kant and others had hypothesized, the thickening of matter in the nebula gave birth to the sun and the whole range of planets that, moving in a disciplined manner on the same plane and in the same direction, orbit around it.

The universe around us is extremely cold, on average three degrees above absolute zero (i.e., minus 270 degrees Celsius). A fortunate combination of physical conditions is needed to maintain a comfortable temperature on the surface of a wandering rock in this frozen cosmos for 3 billion years. Not only that, the sun happens to rotate at just the right speed, its magnetic field is not too strong, and there was enough fuel (hydrogen) in its core at birth to last about 10 billion years. The perfect star in the perfect place in the perfect cloud.

But are you sure? This idea of "rightness"—that is, having neither too much nor too little of everything we need—points our minds in the wrong direction. It is, in fact, a retrospective judgment because here we are now, on earth, admiring the starry sky in wonder and scientifically reconstructing the history of the universe. Yet hindsight acts as an enemy when trying to understand evolution because it tends to underestimate all the countless alternative outcomes and scenarios that could have taken place under the same conditions. Hindsight makes anomalies seem necessary and complete, and therefore perfect even though they are not. It even induces us to turn reality upside down.

Looking back over a contingent history, our mind tends to reason in the fatalistic terms of fate and design, selecting some events and ignoring others. It is as if there has never really been a choice. As if everything had already been written in the cards dealt in the opening hand at the beginning. As if necessity had always woven its web. The inevitability of the outcome is a comforting error of hindsight, which makes us retrospectively link the causes and effects, the before and after, the intentions and consequences.

The problem lies in the fact that our minds lead us to reason in the following way: the many cosmic and personal coincidences that had to take place for me to be here at this moment cannot be the result of chance; it was destined to happen. Many studies on our brains have confirmed the strong psychological tendency of *Homo sapiens* toward animism and teleology—that is, thinking in terms of purpose and preferring narratives in which intentional agents exhibit their goals and try to achieve them. Consequently, we like, and find it easier, to think that cosmic and biological evolution moves from the imperfect to the perfect, from the simple to the complex, from inorganic to thinking matter.

By conceptualizing in this way, we eliminate from our awareness the power of critical moments, those subtle imperfections and breaks in symmetry on which the course of subsequent events depends. If, however, we were to make an effort to understand evolution by identifying and relating to the possibilities that existed at a given moment in history—and if we then looked both backward and forward, always starting from the potentials of that specific moment—the many counterfutures (i.e., the counterpresents of today) that filled the past in its critical points would open up before our very eyes. Not only would we see the present as it is actually taking place, justifying it as necessary, predetermined, "natural," and even inevitable in the light of the past, but we would also appreciate the beauty of all the possible stories that have not been realized. These possible but unrealized presents are contingent (i.e., causally dependent) on past events that did not take place. They are what philosophers call counterfactuals, alternative and plausible versions of the past in which a change in critical bifurcations led to a different outcome from the one that actually occurred.

Hindsight is a poison. Let's get rid of it, and the future will seem more open.

COSMIC RICOCHETS

The history of our planetary home, too, has always been subject to alternatives; today we could be living in counterpresents quite different from our own, which seems special to us as it is the only one that actually happened. According to those who have been lucky enough to admire the planet with their own eyes from the window of a spacecraft, the earth seen from space inspires a profound sense of fragility and responsibility toward it. It looks truly unique (and as far as we know, it is). In other words, it seems perfect for life and for us. But for the world to arrive at that point required a planetary career full of drama and contingencies. The truly perfect planet is in equilibrium—and is therefore dead.

This scenario, though, does not represent our case. In the protoplanetary disk of the weak and newly formed sun, there was a cascade of collisions, explosions, and aggregations, culminating in the accidental survival of a few larger protoplanets: a gas giant at a distance, Jupiter; a few outlying ice giants; five earthlike inner planets (Mercury, Venus, Earth, Theia, and Mars); and debris and asteroids scattered everywhere. The sun's third planet at that time was a glowing mass of iron, oxygen, silicon, magnesium, calcium, aluminium, nickel, and little else. By about 4.6 billion years ago, it had formed into a core, a mantle, a crust, and a poisonous atmosphere. The movements of molten iron in its outer core generated a magnetic field that protected it from much of the solar wind. This was an interesting quality, but the enormous volcanic eruptions and asteroid impacts still made Earth a hellish place. Any passing alien scientist at that time would not have bet a penny on its fertility.

Ninety million years later, however, the unthinkable happened. Too close to Earth, the small planet Theia was being constantly disturbed by other impacts in a great random cosmic billiard game. It finally crashed obliquely into its larger neighbor, devastating it and ripping its atmosphere apart. Part of Theia merged and became part of our planet, while another section was dispersed into space, and yet another became the moon. Earth, which was now tilted on its axis, acquired its own satellite, and seasons and tides were born. In 100 million years, an atmosphere re-formed, water collected in a primordial ocean, the complex process of plate tectonics was triggered, and the greenhouse effect began to stabilize the climate. And so Earth started its career as a living, dynamic, evolving planet. At this point, the conditions needed to host living organisms were perhaps already virtually in place, but it was too soon. And the solar system, which is not as stable as it seems, had other brutal surprises in store.

Around 3.8 billion years ago, the gravitational resonance between Jupiter and Saturn made their orbits more erratic, provoking a major disruption of Neptune's trajectory, pushing it outward past Uranus and into the center of the Kuiper belt. The Jovian upheaval catapulted icy comets and rampaging asteroids across the solar system, bombarding the inner planets including Earth and the moon. All hell broke loose, with catastrophic impacts that led to an immense quantity of external material falling onto the surface of our planet, which likely included high levels of water, carbon, amino acids, and other organic molecules.

In short, nowhere was it written that Earth was the right planet in the right place and therefore perfect. It is through hindsight that we make the prospective mistake of feeling blessed by a benevolent and inevitable destiny. In reality, the wandering stone that would later become our home earned its title in the field. It only became habitable later, at the end of a geologically turbulent period, when the cosmic collisions taking place in the solar system finally came to an end. We earned diversity through this loss of perfection and uniqueness, and as a result, everything became more interesting.

If our universe was in fact born in this explosive manner, with nothing special to make it unique in any way except for that random fluctuation based on inflatons, who knows how many other exponential explosions of space-time have lit up the supercosmos, populating it with alternative universes, some of them full of life and others not, and some of which, who knows, are adjacent to ours but separate? Perhaps we are in company and will never know it. The only one that fits us well is also the only one we know. Imperfection and contingency do not imply a supreme and unrepeatable improbability—that is, the fact that the universe is as it is by virtue of a series of breaks in symmetry, and that Earth can host life due to a lucky mix of favorable conditions, in no way means that this sequence of critical moments happened once only and thus that our history is unique or "special."

Just consider the thousands of extrasolar planets we are discovering, many of which orbit in the habitable zone—that band of suitableness where their star's brightness and orbital radius virtually guarantee temperatures that could sustain life. It has been estimated that there may be more than 10 billion earthlike exoplanets in the Milky Way. Ten billion in our small cosmic circle alone! Ten billion local possibilities in the darkest of nights. Naturally, being situated in the habitable zone is not enough because you also need liquid water and organic compounds to start with. Yet given the billions and billions of warm, humid planets with surfaces where the process can be tested—that is, with billions of attempts available—it becomes rather unreasonable (as well as presumptuous) to think that abiogenesis, the emergence of self-replicating life-forms from the chemistry of inanimate matter, has only succeeded on our rounded boulder of rock and metal lost somewhere out in Orion's belt.

Sharing this peculiar eventuality with other planetary systems does not mean that history has followed the same path everywhere, and nor does it mean that our mix of amino acids, nucleotides, sugars, and fats was the perfect recipe. Our chemical dish consisted mainly of carbon, nitrogen, oxygen, and hydrogen, with a splattering of phosphorus, sulfur, and iron. Elsewhere, who knows? The alien bacteria we may well discover in the near future will provide us with the answer. We will be able to compare them to our own microorganisms that are capable of living in extreme conditions, study what survival routes they have found, and understand just how wide the range of diversity that can be explored is. As far as we know, the first steps of life on earth also crossed many critical moments. And many times we came close to failing, but never completely.

And here begins the natural history of biological imperfection, the heart of our journey.

2

IMPERFECT EVOLUTION

> He proved admirably that there is no effect without a cause, and that, in this best of all possible worlds, the Baron's castle was the most beautiful of castles and his lady the best of all baronesses.
> —Voltaire, *Candide, or Optimism*

If you are born, you eat, grow, interact with the environment and with others, possibly reproduce, and then die; and if, together with many other similar dying people, you belong to a population that evolves over time, then you are alive. This list of unique faculties evidences the imperfection and paradoxical fragility of the wonder we call life. Like a never-ending relay race, countless individual existences appear and disappear, are born, and end and sacrifice themselves so that the species may continue for a while longer. If this is the best of all possible worlds, it is better not to know the others.

THE MOST INGENIOUS OF IMPERFECTIONS

We still don't know whether or not the beginning took place in a bubbling of random interactions in the shallow waters bathed by the sun or in the dark ripples at the bottom of the oceans near hot hydrothermal springs, under overwhelming pressure.

Maybe, in the pores of these sulfurous chimneys, lying on the border between boiling and freezing water, the first membranes closed in on themselves, trapping the essential ingredients for life that were already present within tiny bubbles of water. Those membranes were not impermeable, and they exchanged materials with the outside world, nutrients coming in and waste going out. Then, one fine day around 3.5 billion years ago, one of these bubbles created a primitive metabolism, growing larger and larger, and then splitting into two. It had learned to duplicate itself. Inside its offspring, a substance also replicated itself, forming a necklace of nucleotides that doubled up and made a mold of itself, acquiring the ability to transmit the information needed to build the organism. And so, the game began, a potentially endless game of self-replication.

Once again, having passed the critical moment, a self-nourishing mechanism triggered an endless process. RNA chains gained the ability to make complementary copies of themselves; those copies then separated and replicated themselves. They could also take on different spatial and functional configurations, performing operations on other molecules such as separating, cutting, and aggregating. The RNA chains multiplied, diversified, and became a population. The randomly more efficient versions prevailed over the others. During this molecular evolution, a new polymer, DNA, entered the stage. Being more stable and therefore longer, it was an improvement on RNA and was more efficient in replicating itself thanks to its two complementary, paired strands.

The structure of DNA is incredibly compact due to a coiling double helix with a twist along its central axis that is followed by four more coils. At this point, the web of nascent terrestrial life became an interaction between three biopolymers, swimming in the restricted but membrane-protected water space: the DNA commander, a reliable replicant; the RNA ambassador, which was a less stable replicant, but an indispensable intermediary between the genome and proteins as well as being a regulator of processes; and the amino acids, which were more operational nonreplicants, with countless three-dimensional spatial conformations that performed functional operations.

The first replicants were primordial protobionts equipped with only RNA. It is from these that the fearsome and ancient viruses probably evolved (inconveniently, *they* really border on lethal perfection, as we unfortunately saw with the SARS-CoV-2 pandemic), at a very early stage of evolution preceding the appearance of the universal common ancestor of all living things that consisted of a single cell, the smallest living structure. Thanks to molecular selection, after who knows how many billions of failed attempts, protein synthesis finally started, in an incredible circular process in which certain proteins (enzymes) oversee the process of producing other proteins. Which were born first? It is difficult to say.

Nothing like it had ever been seen before the appearance of this somewhat byzantine but amazing biochemical machine: a portion of the double strand of the DNA (a gene) unpairs itself; one of these unpaired strands produces a copy of a complementary RNA strand using the enzyme polymerase; and this transcribed strand (messenger RNA) is "scanned" by ribosomes and produces another copy that consists of a chain of amino acids, via the genetic code of redundant correspondences between the sixty-four RNA triplets and the twenty amino acids. Finally, the RNA-translated chains become proteins and adopt the elaborate spatial conformations that make all vital processes work by following the instructions contained in the initial DNA. An engineering masterpiece? Yes. Perfect? By no means. If it were, it wouldn't work.

This is where the most ingenious of imperfections comes into play. From one duplication to the next, DNA is faithfully transmitted, but in the process there are random copying errors. Duplication is always imperfect, with slight variations, deviations, and recombinations. DNA has this crucial ambivalence: it is stable, for otherwise there would be no transmission of genetic information; but it is also variable, for otherwise there would be no evolution. Errors in evolution are generative; they are the lifeblood of change. Mutations are deletions, substitutions, or additions of the small letters, the four nitrogenous bases that make up DNA.

If mutations occur in germ cells, they become hereditary and are passed onto subsequent generations. Yet identical reproduction, through cloning over and over again, is worthless by itself. To survive in a changing environment, you must be able to vary. Random mutations mean that each individual is a carrier of unique differences, and has a greater or lesser chance of surviving and reproducing in a given environment. Natural selection is a filter that feeds on chance, causing populations of organisms to evolve. It is fueled by slight innovative imperfections, those deviations from the average, and by the generational disobedience that each individual carries with them from birth.

These mutations must not exceed a certain number, however, because most of them are either neutral (they make no difference for the individual organism) or harmful. Mutation is therefore a two-faced Janus, indispensable but imperfect; it has a positive side as it is the driver of evolution and diversity, but it also has a negative side because, through mutations, a cell can lose control and become, for example, cancerous. And, like Janus, the god who waits in the doorway, it looks to the future and past at the same time.

THE GREAT COMPROMISE OF MULTICELLULARITY

This imperfect and wonderful machine, which presses and grinds random variations like mills do with wheat, produces a level of diversity that is ubiquitous on earth. There are 250 different cell types in the human body alone, tens of thousands of different proteins, and millions of species that have diversified starting from the first bacteria, from prokaryotes to unicellular eukaryotes (the result of symbiosis between prokaryotes) and then on to multicellular organisms with differentiated cells, such as fungi, plants, and we animals, who are the outer reaches of the entire biodiversity empire. Yet all the evidence we have seems to suggest that even this astonishing diversity is only a minimal subset of what is potentially possible. Right from the outset, there were many more potential combinations, with only a few solutions being explored. Of all the combinations that have actually been realized, one must remember that 99.9 percent of the species that existed in the last 3.5 billion years on earth have already become extinct (a huge waste), with most of them vanishing before we appeared, although the self-styled *Homo sapiens* is now adding to this list, having already exterminated more than a third of all other life-forms in the last five centuries.

Sometime around 2 billion years ago, a cell underwent one of its many random mutations, and when it duplicated itself, the two sibling cells failed to separate and remained stuck together. At that point, a further mutation differentiated the two sibling cells so that they performed a slightly different function. Thus the engine of evolution set off again: variation and differential survival. Those two cells have a small advantage over the others: they stick together and share the workload of survival. If that advantage is hereditary and is selected—that is, if the environmental pressure favors it—then the two cells will continue living together, although to replicate they will separate, each one retaining the mutation of the other and thereby adding complexity to both.

At this point, the cells are part of the same system and cannot act alone. Not only do they have to cooperate with each other, they also have to differentiate in the right way, and so for the first time they have to accept being controlled from above. They all possess the same genome, but they are regulated and expressed differently to perform different functions. They are now part of a multicellular organism, our body, an even larger and more complex community than a single cell and an organized and coordinated community. From here, from the first fertilized cell onward, the DNA conductor was able to perform another marvel: some genes became architects and began to oversee the development of the different body plans (being a mollusk, crustacean, insect, vertebrate, and so on). They separated the various tissues and placed them in the correct position. As site managers, from that day on they directed the construction of the various body parts (limbs, antennae, segments, heads, and tails) and dictated symmetry (front and back, top and bottom, left and right, entrance and exit). And so it was that, starting from a limited tool kit (the developmental genes), nature, via DNA, produced the extraordinary diversity of body plans of millions and millions of different multicellular species. At that point, organisms began to eat each other, dead or alive, and developed a taste for it.

But on this occasion, too, there was a price to pay. If a cell lives alone, committing suicide will never cross its mind. The imperative in life is to resist, multiply, and occupy the space available. If it is damaged, it is perfectly okay to survive in the damaged form. At worst, it will have fewer descendants. But in the great metropolis of a multicellular organism, if one cell is damaged by mutation, radiation, or trauma, it will harm the rest of the organism and must therefore be prevented from living. The collective prevails over the egoism of individuals and parts. The cell, willingly or unwillingly, must sacrifice itself for the common good.

A crucial program of cell suicide—apoptosis, meaning programmed cell death—has developed in the course of evolution. It is written in the DNA of each of our cells, and certainly gave a great selective advantage to its first multicellular carriers. One of the consequences of multicellularity and the development of bodies is that cells have to function in unison with all the others in a form of controlled police state. The immune system monitors, and each body is a sculpture that renews itself every day by producing billions of new cells and sculpting itself by removing billions of others. Whenever a cell malfunctions and becomes dangerous for the whole body, apoptosis is automatically triggered and the cell dies. But things do not always work out for the best.

The multicellular condition is an imperfect compromise between cellular egoism and organism-level cooperation. In evolution, when systems with several coordinated parts develop, "free riders" will sooner or later emerge that enjoy the help of others around them but give nothing in return to the common good. Acting in this way, they gain a strong Darwinian advantage because they benefit from the cooperation of others even as they contribute nothing themselves. They are similar to tax evaders, who pay no taxes but then go to the public hospital when they need to. This is why the body has evolved its own internal police force. When a selfish cell manages to evade controls and reverts to a single-celled logic, it becomes a problem.

The disobedient ones that escape the imperative of apoptosis are the cancer cells. Due to accidental DNA copying errors that have escaped controls, environmental influences, or inherited mutations, they become free riders that duplicate indefinitely and dominate the environments surrounding them, absorbing resources like parasites. They are traitors that infiltrate and undermine the cellular cooperation that guarantees an organism's health. The most dangerous tumors have an evolutionary strategy. They derive from genetic instability and environments that favor them due to acidosis, hypoxia, and inflammation. They mutate extremely rapidly, evading attacks and resisting sieges. They diversify, branching out like trees, like species. They corrupt the police force of the immune system and co-opt it to work for their benefit.

This makes no sense. Only imperfect evolution can explain cancer. By killing the organism, the tumor eventually kills its own ecological niche, its host, and inevitably it is game over. Its rampant selfishness is pure foolishness in evolutionary terms—violent and completely mad, eventually dying together with its host. It does not only affect older people who are no longer of reproductive age, in whom cellular aging, genetic instability, and the lack of protection provided by natural selection favor the emergence of malignant mutations. It also affects everyone else, albeit less frequently, and unfortunately that includes children. This makes no sense at all unless we accept that evolution is not perfect but is rather the result of unstable and precarious compromises, which generally work—but not always. In this case, it can be considered a compromise between the primordial single-celled level and the multicellular system. Cancer is a malevolent legacy that goes back to the days when cells did everything themselves: the body's control system is imperfect, and occasionally single-celled selfishness reappears with a vengeance. This struggle between bodies and tumors is therefore an ancient battle between two stages of evolution—an endless battle that has gone on for hundreds of millions of years.

FROM THE POINT OF VIEW OF A MICROBE

To understand just how powerful and resilient single-celled life can be, try looking at evolution from the point of view of microbes. It will amaze you. The "animalcules" observed by microscopist Antonie van Leeuwenhoek in the second half of the seventeenth century are tiny, self-replicating, single-celled organisms that exchange genes with each other and can live literally anywhere. If life exists on other bodies in the solar system or distant exoplanets, it is likely to look like these microbes, and not like little bipedal green people with brains. The complicated, imperfect, but fully functioning biological nanomachines invented by microorganisms, and later inherited by plants and animals, have largely shaped the biogeochemical cycles of the earth, and perhaps one day we will discover that they have done much the same elsewhere.

The ubiquity of microbes is well known; most of today's terrestrial biodiversity consists of bacteria, and all life on our planet is descended, as already mentioned, from a common microbial ancestor that lived at least 3.5 billion years ago. Thus the fact that the animals we are familiar with appeared 600 million years ago means that 85 percent of the biological evolution on earth was totally microbes, archaebacteria, bacteria, and little else. This comes as a shock to anyone still convinced that natural history on earth was inexorably dedicated to developing complexity and intelligence. It all depends, however, on what we mean by intelligence and complexity; the bathyscaphes that have been combing the hydrothermal vents on ocean ridges, the endless abyssal plains, and the oceans' muddy floors for decades have now discovered a world of bizarre creatures that were previously unimaginable. The darkness is interrupted by flashes of bioluminescence, and the universe of sponges, mollusks, crustaceans, giant worms, starfish, fish, and foraminifera has little to envy in the teeming life of a rain forest.

Still, our debt to microbes and the seemingly simple side of life stretches well beyond the fact that they have long preceded us (Falkowski 2015). Approximately 2.4 billion years ago, if not even earlier, cyanobacteria learned to convert solar energy into organic compounds (whereas terrestrial plants have "only" been doing this for 450 million years). Through oxygenic photosynthesis, light energy was used to break down water, obtain hydrogen, and absorb carbon dioxide to produce organic matter, releasing oxygen as a waste product. As an entirely contingent side effect, a previously absent, highly reactive, and toxic gas that promotes combustion began to spread throughout the atmosphere: oxygen. For us, oxygen means life, but this was not always the case.

The event triggered a catastrophic revolution: the Great Oxidation Event. Over hundreds of millions of years, the atmosphere filled with oxygen, eventually arriving at the 21 percent that has been stable (fortunately for us) for at least the last 800,000 years. The ozone layer, which protects us from ultraviolet rays, was also formed. The planet cooled dramatically due to a reduction in atmospheric methane, and the earth turned into a giant snowball for 300 million years. The life-forms that until that moment had proliferated in anoxic conditions were wiped out because for many of them oxygen was a poison, and the few survivors were pushed back into marginal niches (today we can find them, for example, in our intestines and those of ruminants). Instead, other microbes and then animals learned to breathe in oxygen, releasing water and carbon dioxide as waste that is then reabsorbed by phytoplankton and plants. In this great intertwining of biogeochemical regulatory cycles, the microbes that survived these climate fluctuations and the phytoplankton created the conditions that keep us alive today; basically, it was microbes that made the earth a habitable place for us to live in.

There is another reason why we wouldn't be here without bacteria: the first eukaryotic cells arose from symbiotic associations between microbes around 2.7 billion years ago. The principle is that, while you cannot create something from nothing, you can recombine what already exists. Some bacteria were absorbed and incorporated into other bacteria, forming more complex cells with internal organelles and a nucleus to protect the DNA from contamination. These structures included the mitochondria and chloroplasts, the energy generators of animal and plant cells, which are in origin bacteria that had been ingested and maintained as symbionts, and retained their original DNA. This was an early example of cooperation between cells that would later become successful and be repeated in the evolution of the first multicellular colonies. So every single one of our cells hosts a former bacterium that has given up its independence.

This is how we continue to live with bacteria today, fluctuating between collaboration and belligerence. We transport billions of them on our skin, in our mouths, and in our noses. Our intestines contain a rich zoo of bacteria, archaebacteria, viruses, and fungi, more than ten thousand species (mostly unknown), in a microbial ecosystem acquired largely after birth from the parent and environment, unique to each individual but typical of each species. This "microbiota" in our bodies enables us to digest an amazing variety of substances, and it has a fully working metabolism and healthy immune system. Surprisingly, it has recently been discovered that when it malfunctions, it may even play a role in diabetes and some neurodegenerative diseases.

In short, our bodies are crowded apartment blocks that house billions of microbes in an unstable state of equilibrium. Given that nature is always ambivalent and never perfect, some help us live (they are symbionts: we need them, and they need us), while others exploit us without harming us too much (they are harmless parasites or diners), and yet others infect us and make us feel ill (they are pathogens). Furthermore, some start out as harmless but later become pathogens. And there are many others whose functions we have yet to discover. What's more, we are not alone in our bodies. We are "holobionts." For bacteria, after all, we mammals are simply an excellent habitat and vector, and in evolution, we must also look at the world from the point of view of others. Diet, hygiene, and our lack of contact with nature are reducing the biodiversity of the human microbiota in industrialized countries to the point that some scientists are suggesting we should build a large Noah's ark, a global biobank, to preserve the gut microbes of native peoples, which are far more biodiverse than ours (Dominguez Bello et al. 2018).

Even if this hurts our anthropocentrism (and also animal centrism), we can notice considerable asymmetry between the bacteria and ourselves: without their massive biomass, we could not exist, while they have no need for our noisy assemblies of different cell types, tissues, and organs. They live in the most severe conditions, in the absence of oxygen and light, at extremely high temperatures and pressures, in salt and inside rocks. If there is no solar energy, they opt for chemosynthesis using inorganic substances. Were they to travel round the solar system, they would do well on Mars, in the hidden ocean of Europa, and on the dunes of Titan. We have learned to keep them at bay with antibiotics (thereby depleting the microbiota with unpredictable consequences), but they mutate quickly and become antibiotic resistant. The next time you consider grandeur and perfection, think of microbes; they were here before we were, they have chemically transformed the planet, we could not live without them, and everything suggests that they will continue to dominate the planet long after the demise of *Homo sapiens*. As microbiologist John L. Ingraham (2012) put it, microbes are our progenitors, inventors, and preservers. Some of them sometimes become our opponents and kill us, and after a devastating pandemic we know exactly what that means. But we must remember that it is we who are the recent intruders in their long-established and developed world.

SEX AND OTHER CATASTROPHES

Microbes are also known for having invented another hellish danger: sex. If we want to increase the chances of genetic variation, in addition to normal mutations, we can mix the genes of two organisms. The sexual act of microbes is both efficient and sensible: they directly exchange genes horizontally through cytoplasmic bridges and other recombination techniques. We multicellular beings, on the other hand, have chosen a much riskier route, which exposes us to myriad imperfections.

Sex, fun though it is, has a cost. Courtship, resistance, mating, and parental care require considerable energy, which must be diverted from other vital activities such as feeding oneself and avoiding predators. A large number of species, in fact, have gotten away from sex as quickly as they could; for instance, plants can bud from a cutting, many reptiles give birth through parthenogenesis, and others clone themselves. These are all quick, painless, and cheap solutions, but they have one major drawback: they do not generate diversity. The parent is identical to the child, which is anything but ideal. So DNA invented the distinct sexes.

Females and males mix and recombine their genomes to give birth to a child who resembles the parents, but who is never totally identical, and this means that no child is ever the same as another. If you remember, variation is the fuel of evolution. So we can assume that at one time, at the dawn of multicellular life, sexual and asexual reproduction competed as two alternative strategies. Those whose offspring were all clones found reproduction easy, but were exposed to the risk of disease. If a pathogen struck one of them, it struck them all, and there would be no more offspring. On the other hand, those who opted for sex, and thereby generated offspring that were all genetically different from themselves and each other, spent much more energy, but with a big plus: if the disease affected some of their children, some of them would probably be resistant or immune due to their genetic diversity. Therefore their lineage was safe.

Today we have clear evidence of this in species that have the two options. Among them, asexual reproduction prevails in stable situations, but when there is an external threat or environmental stress, sexual reproduction reappears. Sex is thus the stratagem that DNA uses to produce diversity with each generation, and diversity in nature is an insurance policy for life and the future. The more diverse a biological population is, the healthier it is because it can better resist the attacks of pathogens. If, however, a clone acquires a negative mutation, it carries it forever while accumulating others with the passage of time. Sex is a natural vaccine that has protected us from viruses and bacteria for six hundred million years. The moral of the story is that doing such things as reducing diversity, homogenizing and standardizing the world, raising animals that are all clones, cultivating monocultures, speaking the same language, and thinking in the same way is never a good idea.

Sex, though, has many rather imperfect drawbacks. You can no longer happily reproduce on your own; rather, you are dependent on individuals of another sex, with whom it is not always easy to interact (try asking the male spotted hyena that has to relate with bigger and more muscular mates, which are rather cantankerous, and with a disturbing false penis in between their hind legs). Apart from being expensive and dangerous during the physical act, sex also generates a cascade of asymmetries between males and females. The latter usually produce a limited number of sex cells, which are nutritious and valuable, and hence must not be wasted. The former, conversely, ejaculate millions of sex cells, most of which disappear without a trace. Moreover, females generally bear most of the reproductive costs and maintenance of their offspring, whereas males, with few exceptions, invest little in the care of their offspring.

As a result, females tend to choose their male partner while males have to find strategies to be chosen, either through fighting with other males or by seducing females with gifts and forms of exhibitionism. The costlier and more elaborate the male effort is—that is, the more it is a handicap for the male suitor—the better the chances of winning over the female because, in this way, the future mother has proof that the male is healthy and strong (for if he is still alive after all that effort, he is either lucky or strong). This means that in nature, we males are often ridiculous and indeed the champions of imperfection. It is likely that evolution encouraged sex by making it particularly pleasurable, precisely in order to compensate for all of these dangers and defects.

THE CHAMPIONS OF IMPERFECTION

The sample of male horrors found in nature ranges from the parasitic partners of the monkfish—a fish from the abysses whose gigantic females carry around a dozen tiny males glued to their skin, sucking out every internal organ except the gonads and making them practically walking testicles (after all, when you live in the dark, immense ocean and you meet a partner, you would be better off holding on tight . . . even too tight)—to the males of mantises and various spiders that are devoured during or immediately after mating (rather like a bad wedding gift). And we can also think of male peacocks with their exaggerated tails or deer with their bulky, sumptuous antlers.

Human males are genetically imperfect, too. While the chromosomes of females are paired and symmetrical, including their sex chromosomes, males have an unpaired sex chromosome, Y, that does not recombine (and does not renew) with X. Some biologists argue that it is gradually wasting away due to a lack of mutations. We human males nevertheless risked extinction even earlier, and in the most ridiculous of ways. In fact, human genetic data show that somewhere between seven and five millennia ago, the human Y chromosome went through a rather dramatic "bottleneck" that reduced its variability. This means that in many human populations in Eurasia and Africa, the number of humans plummeted around that time. It has been estimated that population levels halved from twenty to ten million. What could have provoked such a mass slaughter of reproductive males?

In evolution, genetic bottlenecks are usually due to profound ecological disturbances that drastically reduce population size, but there have been no known environmental catastrophes that have been so selective that only males were killed off. Perhaps it was due to an epidemic of a mysterious uniparental disease, or to an explosion of infant mortality in males? Or possibly in agropastoral societies, social inequalities spread, and with them polygyny, with a small number of men reproducing with a large number of women? Or perhaps the bottleneck was due to the fact that in the Neolithic period, founding groups of farmers and shepherds were composed mainly of men who had migrated several times?

Recently, by combining anthropological, archaeological, and genetic data together with mathematical models, some experts have suggested a different and more disturbing scenario: the formation of patrilineal parental groups (genetically homogeneous internally) that were in strong competition with each other, resulting in the survival of only a few, and the consequential massacre of males in battle and indeed entire tribes with all of their Y chromosomes. In practice, males risked extinction by killing each other and taking no prisoners in rival groups (Zeng, Aw, and Feldman 2018). Very *sapiens* behavior, indeed.

Were this the real scenario, it would mean that, for thousands of years at least, the human male has had serious problems managing his testosterone levels in society, so much so that he created a dangerous genetic bottleneck through massacres and absurd generalized wars between clans to conquer a piece of land (or the women of others). Hurting himself and dragging the rest of the world down with him, and strutting around like a peacock while doing it, seems to be a rather deep-rooted evolutionary prerogative in the human male, such an imperfect creature that he just about managed to extinguish himself.

It must be said, in partial defense of the eternally insecure males, that they have been through hell for at least a couple million years. It is not possible to know when our females are ovulating, which is a rare phenomenon in primates, so we males are never sure when our partners are fertile, and more important, if we really are the parents of the offspring. All we can do is to watch over them night and day, living in permanent fear of betrayal, which is common in many other species, even the most unsuspected. In many birds, such as swans, males and females are identical in outward appearance. The males are not in competition to be chosen by the females because the pairs are perfectly monogamous. They choose each other and stay together all of their lives. Or so we thought. In fact, when scientists carried out the first genetic tests on eggs, they discovered that there were almost always illegitimate offspring in the brood—the result of "extramarital" encounters. So even monogamy is not perfect. After all, evolution has shown us that a touch of infidelity increases genetic variability, which we must unfortunately admit could be a good thing in evolutionary terms.

A WORLD OF POSSIBILITIES

Not only does imperfection teach us to see things as they are, but it suggests how they could be. Having passed the 600-million-year mark, and having put into practice the inventions of the previous two particularly boring billion years (photosynthesis, cells with nuclei, multicellular organisms, cell death, and sex), the production of biodiversity began to accelerate, without paying much attention to morphological perfection. Once the most severe glaciations had passed by and oxygen levels had risen, multicellular organisms liberated their developmental genes and started experimenting with the most bizarre body configurations. They began with the explosion of the mysterious flattened and "arboreal" fauna of the Ediacaran, in which the ancestors of sponges, jellyfish, and corals perhaps emerged, while many other forms became extinct forever. There followed the explosion of the

Cambrian fauna, the strangest ever to have appeared on earth, populated by prey and predators with extravagant combinations of body forms (eyes, mouths, heads, anuses, limbs, antennae, claws, armor, bilateral symmetries, and segmentations). A caravansary that looked very much like something from a science fiction film.

The Cambrian seas of 520 million years ago were already teeming with a multitude of species including the extinct ancestors and distant uncles of all the major animal groups still living on the planet. We will never know which of them was traveling down an evolutionary cul-de-sac, and which was destined to become the ancestor of insects, crustaceans, spiders, or vertebrates such as those anomalous lobed-finned fish that much later, at the end of the Devonian some 370 million years ago, timidly and randomly began to adapt to land. No sooner did an ecological opportunity arrive than life immediately started to develop and explore a wide range of possibilities, only some of which have survived.

The construction principle of these early animals, if we really want to find one, was not engineering optimization but rather experimentation with possibilities and functional specialization in feeding. They lived in a free and permissive world. Unlike in later times, they could indulge in differing forms and solutions. Imperfect evolution at this point began to alternate between periods of diversification (radiation), flourishing stability (such as in the Carboniferous), drastic species reduction (mass extinctions), with the latter always being triggered by large-scale ecological changes that had little to do with the organisms' adaptive capacities, such as colossal volcanic eruptions, asteroid impacts, continental drifts, and climatic oscillations. As a planet, the earth shows no mercy and can periodically deal a tremendous blow to the fragile life-forms that inhabit it.

On each occasion, however, that fragility transformed into extraordinary resilience. Following a mass extinction, the few survivors, which were by no means predestined, occupied the vacant ecological niches and became more diverse than before. Not only is life imperfect, but it is stubborn. This stubbornness expressed itself after the mother of all extinctions at the end of the Permian, 252 million years ago, when 90 percent of all species became extinct due to huge eruptions of molten basalt. Yet from the few survivors (including unlikely animals such as the clumsy mammalian reptile called *Lystrosaurus*, on which no objective bookmaker would have bet a penny), the carousel of biodiversity slowly started to turn again.

In mass extinctions, the fittest do not necessarily survive. The event is too sudden, and there is no time. Sometimes the most adaptable survives, such as the generalist that has a varied diet and accepts different environments. On other occasions, it is simply the luckiest that survive—those that happened to find themselves in the right place at the right time. The extinction, 202 million years ago, at the end of the Triassic, which exterminated a large part of the huge monstrous reptiles of the period, became the great opportunity for the slow and difficult rise of the dinosaurs in the Jurassic, which then dominated the Cretaceous until 66 million years ago. The lucky survivors of the previous catastrophe become the victims of the next one.

For half a billion years, explosive radiation and similar phenomena, together with dramatic ecological crises over and over again, have characterized the evolution of life on an active and unpredictable planet. Imperfect life, therefore creative, on an imperfect planet, therefore fertile. The rest is "recent" history (on a biological scale). Being small, highly reproductive, generalist, mobile, warm-blooded (i.e., able to regulate body temperature internally), and nocturnal, with a habit of stashing away seeds for the winter, does not necessarily mean perfection (so much so that for a hundred million years, these little animals remained relegated to the nooks and crannies left free by the dinosaurs). But through a stroke of good luck, it meant being the perfect animal able to survive the long night of the Cretaceous period that wiped out all nonavian dinosaurs. The dinosaurs are still with us today in the form of ten thousand species of birds, but the ecological niches for large vertebrates now belong to mammals. After the impact, the critical moment of all critical moments, came the era of mammals, marsupials and placentals, small and giant, adapted to land, air, and water. We are children coming from the end of the world of the others. Another clinamen, another cosmic ricochet, without which we hominins would not be here today talking about it.

But before we jump to the bipedal, brainy, chatty primate that emerged from Africa only two or three hundred millennia ago together with many other humans, and thereby introduce the most inessential, fortunate, and creative by-product of imperfect evolution, we must first take a look at the scars that surround us, and try to use them to understand the laws that govern nature's ubiquitous incompleteness and suboptimality.

Here, though, a problem arises. Books and nature documentaries are brimming with poetic descriptions of the wonders of the biosphere, and not without reason. Spectacular aerial shots and close-up drone raids show us the moving beauty of the few ecosystems that have survived human bulldozers and chain saws. How could we fail to admire the charm of a high-altitude cloud forest, an architecture of humidity? How could we fail to admire the rhythmic singing of humpback whales? How could we fail to admire the ever-open mouth of the moray eel, with its double pair of internal jaws that suck in and shred prey in an instant? How could we fail to admire the underestimated sociality and sexual idiosyncrasies of velvet worms, the opportunistic strategies of tapeworms that slip into every nook and cranny of our bodies, the resistance of tardigrades to the most extreme conditions, the eye of a dragonfly, the wings of pterosaurs, the gliding of a flying lemur, the electrosensitive nose of the goblin shark, the buoyancy chambers and jet propulsion of the nautilus, the silent flight of the owl over an unsuspecting prey, the immunity of the honey badger to stings and poisons, and the innate elegance of the snow leopard? How could we fail to admire the magnetic sense of direction of the leatherback sea turtle, which crosses the oceans covering thousands and thousands of miles? How could we fail to admire the barnacle's snakelike penis, which is eight times longer than its body and reaches out on the rocks in search of a female (not to mention the penises of other species that are equipped with hooks, spines, and teeth that remove the sperm of their competitors)? And, not for the overly squeamish, how could we fail to admire the sensitivity of spider hairs or neurotoxin of the blue-ringed octopus, which is thousands of times more lethal than cyanide?

The list of magnificence in nature is endless (Henderson 2012). There is no doubt that all of the examples previously discussed represent extremely efficient ways of living, reproducing, and enjoying the split second of existence that is gifted to every living being. Nevertheless, are we really so sure that behind the most admirable works of evolution there is not also an obscure side, a debt to be paid, a scar? The time has come to throw a few grains of sand into the intuitive wheel of natural perfection.

3

IMPERFECTIONS THAT WORK

> The following day they rummaged among the ruins and found provisions, with which they repaired their exhausted strength. After this they joined with others in relieving those inhabitants who had escaped death. Some, whom they had succoured, gave them as good a dinner as they could in such disastrous circumstances; true, the repast was mournful, and the company moistened their bread with tears; but Pangloss consoled them, assuring them that things could not be otherwise.
>
> —Voltaire, *Candide, or Optimism*

The Irish elk, *Megaloceros giganteus*, became extinct nine thousand years ago. Yet it was not an elk and it was not Irish. It was, in fact, a large subarctic deer. Remains of specimens have been found in a vast area ranging from Siberia to North Africa, although perhaps the last population did resist in Ireland, hence its incorrect geographic attribution. The males had exceptionally large, branching antlers. Those antlers could span 3.65 meters (just shy of 12 feet) and were renewed annually. This enormous expenditure of energy had a precise function: to outdo other males in the contest to win over the females. In short, sexual selection, the second selection process (the first being natural selection) proposed by Darwin, in which (mostly) males struggle directly for reproduction (namely, access to females) through ornaments or armaments. The antlers were the "status symbols" of *Megaloceros*, an adaptation to increase their reproductive success by attracting as many females as they could, possibly without even having to fight other males and risking injury. Nevertheless, if these animals were really so well adapted to reproduce, why did they become extinct?

THE IRISH ELK AND THE FIRST TWO LAWS OF IMPERFECTION

With the benefit of hindsight, considering adaptation as a form of functional and optimal perfection risks leading us astray when trying to explain evolutionary processes. Sitting on the sofa watching high-definition nature documentaries can lead us to intuitively presume that the living world is a large, harmonious, and balanced system in which every aspect has a precise meaning and function, and each follows the role assigned to it by the laws of nature. Who has not thought of perfection when seeing a cheetah running? (But they never show you the animal a minute later when it is out of breath and, if it has not bitten its prey on the carotid artery, is lying down in the shade exhausted). The predator plays the predator, the prey plays the prey, and the old trees watch. One pursues and the other runs away, with varying results. Still, were this to be the case, there would be no evolution because everything would have already been accomplished. The spectacle of natural munificence is not the product of millennia of biological peace and wisdom but rather of eras of upheaval, imbalance, and imperfection that have worked out in spite of everything.

In our idealized scenarios, for example, we tend to underestimate the importance of variations and traits generated purely accidentally, such as through genetic drift—that is, processes of completely random changes in gene frequencies along with reductions in variability due to small population sizes or ecological disturbances, sudden bottlenecks (the rapid decrease of a population due, for instance, to ecological disturbances), and the "founder effect" of a small group migrating to found a colony (and transmitting to the colony a random sample of the original genetic variation). Such random evolutionary phenomena have no adaptive value and are completely unrelated to natural selection. They are history, events, and chance, and nothing else.

Naturally, the imperfection and randomness of these traits can never be so strong that they make the survival and reproduction of an organism impossible, because otherwise its genes would not be passed down and that imperfection would simply not exist. In the past, some people have suggested that Irish elk had gotten themselves into serious difficulty, imagining them dying in the woods, trapped in between the branches by their enormous antlers. In other words, they were asking for it. But we have never seen an animal that has actively evolved such damaging structures and behaviors to the point that it becomes extinct (perhaps the only possible example of a "self-threatening" species is paradoxically *Homo sapiens*, but we will look at that in the final chapter). In reality, Irish elk did not become extinct because of their large antlers but despite their large antlers. The end of the last Ice Age triggered a series of rapid climatic changes that were probably unfavorable to the elks' survival, and only at that point did their antlers, which had previously been so useful, become a hindrance due to changes in diet, vegetation, and other environmental circumstances (assuming there was no human involvement in its extinction).

And this is the point. In natural history, one can also end up becoming a victim of one's own previous success. The glorious antlers of the Irish elk were not perfect in themselves, and at some point they proved counterproductive, a costly male luxury. So now, thanks to the moral of the story behind the extinction of the Irish elk, we can outline the first two laws that explain imperfection in nature. The first is the more banal, and states that *in the form of mutations, genetic drift, mass extinctions, contingent, and perhaps rapid ecological upheavals, chance often unpredictably changes the rules of evolution in such a way that a trait that was previously a big advantage and well tuned by natural selection becomes a handicap or a dangerous imperfection*. Therefore something perfect can potentially turn into something imperfect through changes in circumstances that we cannot control. The nonavian dinosaurs perished not because they were in any way inadequate (they were wonderfully well-adapted and diverse social reptiles) but rather because they suddenly found themselves catapulted into an environment that was alien to them. Yet a small proportion survived.

The Irish elk also provides an example of the second law of imperfection in action. Its huge antlers, which grow back every year, were a costly investment for the males, not to mention being a hindrance when moving around. The big advantage in reproductive terms, by sexual selection, was particularly expensive in terms of survival—that is, natural selection. The peacock's tail, the bright colors of many other birds, and a great many other traits that have evolved by sexual selection show that animals frequently have to find a balance between conflicting selective pressures: What should I do, try to become irresistible in the eyes of my females, or escape from predators and try not to starve to death? Given the need to survive yet reproduce too, the solution adopted from time to time can be nothing more than a suboptimal and unstable compromise—that is, an imperfection—that could nonetheless be temporarily useful when necessary. *Imperfection in nature often stems from the need to find a compromise between different interests (e.g., between males and females) and antagonistic selective pressures.*

At this point, imperfection begins to appear scientifically relevant. Suppose you are studying the optimal living conditions of an imaginary sponge that is immersed in a liquid with the nutritional and hydrodynamic characteristics of the ocean environment. Now compare your results to real sponges that are living in the seas (which are among the oldest living organisms and are animals like us, even though it may not seem so), and you will notice that they will more or less tend to have the ideal expected shape. Nevertheless, you will notice considerable morphological diversity between the individuals, populations, and subspecies of sponges. This is because each species is subjected to a wide range of independent selective pressures. If only nutrition is considered, the revealed shape will, in fact, be ideal. If, on the other hand, other problems and environmental requirements (e.g., a stable predator in the vicinity) are superimposed, the resulting shape will be an imperfect compromise between adaptive traits for one function and adaptive traits for another.

Evolution represents an ongoing dilemma. I am a bird and there is my brood, meaning my genetic future. How can I manage to get more food to the nest without spending too long away? The New Zealand kiwi had no mammalian ground predators in the past so it went for running, claws, a superb sense of smell, and nocturnal life. It shrank and stopped flying, and its evolutionary strategy was then to invest its energy in laying huge, disproportionate eggs, each one weighing more than a quarter of the entire animal, from which robust, full-bodied young hatched. When human settlers introduced invasive, egg-eating species to the islands, for the kiwis, it was an absolute disaster.

In all of these cases, living beings do not climb the highest peak of optimal adaptation like an albatross in flight but instead are perfectly happy to settle for a "local maximum." Returning to sponges, this works well if it is true that some can live for up to a thousand years. It is therefore better to discard perfection if you want to survive. If you do not believe it, try asking a sole or plaice, which have had to become abnormally flattened in order to survive: one eye has migrated to the other side, and the fish can spend their entire lives lying on only one side of their body without ever developing bedsores.

THE MARK OF USELESSNESS

Darwin struggled with perfection all of his life for two reasons, which are still relevant today. The first was linked to the saturated and devotion-ridden natural theology texts that he studied at the University of Cambridge, such as philosopher William Paley's 1802 tract that was full to the brim with inspirational and moving descriptions of the wonderful and perfect "adaptations" (a finalist term that preceded evolutionary theory and defined a form that was admirably "apt" to perform a function) of living organisms. According to the design argument, these marvelous and optimal adaptations demonstrated the work of a creative divinity, given that such natural masterpieces of complexity and organization could not possibly have been produced without a purpose or intentional plan.

How could the monkfish have evolved that fake (perfect!) fish that dances in front of its mouth to attract prey, which is in fact a modified dangling lobe of the spine of a dorsal fin? How is it possible that this same adaptation, a mock decoy fish that is identical to certain fish on which it grows its larvae (with the same fins, same tail, and same wriggling), evolved in the bivalve mollusk, *Lampsilis*, from a flap of its incubation sac? How is it possible that marsupial and placental mammals, two alternative ways of being mammals, were faced with similar environmental challenges, but evolved similar structures and behaviors in different parts of the world? Echolocation was invented by bats, but also by some South American birds, the oilbirds. These are natural but misleading questions (in the sense that they assume some inherent trend toward perfection).

In fact, from a creationist perspective, it is entirely predictable that organisms are perfectly equipped and well suited to the environment, which has also been created. Darwin's *The Origin of Species* is also full of descriptions of "fine-tuned adaptations" and coadaptations between organisms. The natural world is full of them because evolution can act on minimal differences in structures and behavior; it is like the imprint of the art of the breeders, but much superior for the refinement of the results (Darwin [1876] 2009). Likewise, today we can admire the beauty and functionality of the molecular structure of hemoglobin, like the spiral shells of nautiluses and ammonites. But Darwin is careful, in two ways, not to confuse his paean to adaptation with those of the natural theologians.

On the one hand, he goes to great lengths to provide a wealth of examples that show how those apparent wonders of biomechanics and physiology (e.g., the admirable crafting of the eye, the surprising mimicry of an insect, and the airy trabeculae of bird bones) can instead gradually evolve through the slow, unintentional scrutiny of natural selection on the random variation of individuals; on the other hand, on numerous occasions in his work, he points out how the norm in nature is imperfection and not perfection. He perceived that the crux of the controversy between evolutionism and fixism might center around the "peculiarities" of nature—that is, the crucial issue of imperfection.

And so the natural history of imperfection was born. For Darwin, morphological and structural evidence represented the most important empirical evidence for evolution. At that time, it was already known that living beings show evident structural "homologies" (e.g., in the limbs of all vertebrates) with superficial successive modifications, as if evolution had selected a limited set of basic morphological patterns and body plans, making only simple variations on the same themes from then on. For Darwin, this could only be explained through genealogy, namely common descent with modification: homologous structures show that all of those animals came from common ancestral forms, which were then acted on by natural selection as environmental conditions changed.

Darwinian evolution thus arises from a dialectic between "unity of type" (i.e., inherited morphological structures) and "conditions of existence" (i.e., external selective pressures). In other words, between inertia and historical constraints on the one hand and contingent environmental situations on the other. It is already clear that the conditions needed for engineering perfection are simply not there. The forelimbs of a human being, a mole, a horse, a dolphin, and a bat are used today for completely different functions (grasping, digging, running, swimming, or flying), but they are all based on the same standard model—that is, the same bones can be found in the same positions. These limbs are not ideal for the needed tasks, but without a doubt they demonstrate common descent.

Natural selection is not omnipotent, nor is it the layperson's substitute for the great designer. It must occasionally make compromises in accordance with the materials available, which are full of internal constraints and physical limits. Selection can only improve organisms with regard to the contingent organic and inorganic conditions of life, and does not strive for impractical perfection. Adaptation thus becomes a relative concept, and the past leaves its mark in the form of imperfections and peculiarities. For Darwin, this is also demonstrated by the many rudimentary or vestigial traits that persist as totally useless remnants in animals. Should environmental conditions change, previously useful organs may become a nuisance, but not up to the point that they are removed. And so, there they remain.

Atrophied eyes (what use are fragile and costly eyes if you have fled to live in a dark cave?), discarded wings in hundreds of species of birds and insects (what use are fragile and costly wings if, as in New Zealand before humans arrived, you have no predators and can find food on the ground?), male breasts, the beginnings of hind limbs and pelvises in the boa, teeth in whale fetuses or the small pelvic bones that remain in adult whales, rudimentary petals, and so on—these are all signs of history, legacies of distant relatives, disused structures that evolution will tolerate for a while or will reuse at a later date should they be needed, as in the case of the eyes that can be found below the skin of some moles, or penguin wings used as fins, and insect wings that are reused as halters. Nature, writes Darwin (1872, 397) in a beautiful passage in *The Origin of Species*, bears the indelible "plain stamp of inutility"—the imprint of imperfect traits, which are "extremely common, or even general, throughout nature."

You will be hard-pressed to find an animal that has no rudimentary or useless traits such as the wings of the New Zealand kiwi. Nature is full of leftovers. In 1867, Darwin's enthusiastic German pupil, zoologist Ernst Haeckel, officially suggested introducing a new term, "dysteleology," to refer to the study of rudimentary, imperfect, functionless organs in the animal and plant worlds as supreme evidence of evolution. Darwin was all in favor, but the recommendation was not followed up and the "science of imperfection" was never formalized.

Nevertheless, the English naturalist in later years did outline a theory of uselessness, mentioning a number of distinct causes: ancestral influences, as in the case of vestigial traits, and casual side effects, as in the case of the sterility of hybrids, such as the mule and hinny, born of parents of two similar but distinct species (donkey and horse)—a totally useless and even negative trait considering that infertility prevents reproduction. According to Darwin, this is the accidental result of differences in the reproductive system of the two species, although uselessness might arise from structural correlations that natural selection fails to eliminate, such as when an adaptation leads to the modification of one part of a structure while generating nonadaptive effects in another part, or it might be generated through the consequences of developmental processes, as is the case in human male nipples, which persist in adults because they develop before sexual differentiation in the embryo. Therefore, according to Darwin, there are principles independent of natural selection that can explain the spread of many useless, imperfect, suboptimal traits.

Evolutionary processes tolerate unpleasant side effects as long as possible, or in some way compensate for them, because once a certain path of physiological transformation has been undertaken and written into the orchestration of the organism's development, it would be far more costly or impossible to go back to the beginning, resetting everything in order to get it right. The genes and developmental systems of fish, amphibians, reptiles, and then mammals took place in sequence, following artisanal and not industrial logic.

One classic example highlights more than any other how bizarre, if not ridiculous, a persisting historical legacy can be. In the giraffe, the recurrent laryngeal nerve is fundamental because it is involved in swallowing and vocalization. But instead of going directly from the brain to the larynx as any engineer would design it, it takes a long and different route. It actually touches the larynx (its final destination), but does not stop, continuing all the way down the neck, following the vagus nerve and passing under the dorsal aorta near the heart. From this point, it returns all the way back up the neck to the larynx after covering almost four meters (the equivalent of about thirteen feet in a human). This makes absolutely no sense at all. It is a clumsy compromise between the recent elongation of the giraffe neck caused by natural selection and the ancient legacy of the vagus nerve in the fish from which we all descend, whether giraffes or humans. The vagus nerve of the fish travels to the gills via the shortest route, while ours goes the long way round (Dawkins 2004; Coyne 2009). Starting all over again would have been impossible, so moving on from one compromise to another, the anatomy of our internal organs with all of their complications and asymmetries has become a tangled jungle.

USEFUL ODDITIES

What if futility were a resource, too? Vestiges are not just useless reminders of the past, such as our goose bumps, which are a legacy of the erection of our hair. Sometimes evolution ingeniously reuses superfluous structures that had evolved at an earlier date. Darwin (1872, 158) writes that "structures thus indirectly gained, although at first of no advantage to a species, may subsequently have been taken advantage of by its modified descendants, under new conditions of life and newly acquired habits." The English naturalist realized, for instance, that the sutures in the skulls of young mammals had not evolved as an "adaptation" to facilitate the exit of the head through the birth canal, as hindsight would lead us to believe. In fact, they are also found in the skulls of young birds and reptiles, which have no need of them given that they simply have to come out of a broken egg. This means that the trait must have evolved early in the common ancestor of reptiles, birds, and mammals, perhaps due to growth constraints, and then only later did it become unexpectedly useful for birth in mammals.

Darwin notes that unnecessary and imperfect traits therefore offer a further advantage to scientists. Since natural selection has not recently acted on these traits by shaping them for adaptive reasons, Darwin (1872, 402) states, using a linguistic metaphor at the end of chapter XIV of *The Origin of Species*, that these rudimentary and negligible organs "may be compared with the letters in a word, still retained in the spelling, but become useless in the pronunciation, but which serve as a clue for its derivation." Essentially, given that natural selection and other mechanisms are not interested in imperfections, they become like archaeological traces, or valuable clues that can be used in the reconstruction of the evolutionary histories of and kinship among living beings. The more insignificant a character is, the more information it provides about lineage relationships between species. Rudimentary teeth and flowers are key features in the classification of animals and plants.

An imperfect, insignificant, or strange trait becomes an almost certain clue regarding common descent when it concerns species that have adapted to different environments. One such example is the external scrotum. Many mammals have one, including us, and we know that it plays an important role in cooling the testicles for the production of spermatozoa. Nonetheless, many mammals—including elephants, hyraxes, anteaters, dugongs, elephant shrews, and golden moles—have testicles inside their bodies. Thus the external scrotum is useful but not essential. So why do these mammals, which are so different in size and lifestyle, not have one? Simply because they inherited its absence from a common African ancestor and managed to cope anyway.

Sometimes peculiarity is a sign of great evolutionary age. Let us solve a riddle. It is a great swimmer, covered in fine fur, and yet it lays eggs. When the eggs hatch, the mother suckles her young from pores on her belly. To early eighteenth-century observers, that duck's beak grafted onto the body of a beaver seemed to be a freak of nature or the rebus of a taxidermist in the mood for a joke. It has webbed feet, but also has claws (the hind ones are poisonous in the male). Insectivorous and nocturnal, it has a vestigial stomach, does not use its eyes underwater, and tracks its prey by means of electroreceptors on its beak. These semi-aquatic animals and their four cousin species of terrestrial echidnas make up the eccentric group of oviparous mammals known as monotremes. Did you guess the answer? It is the platypus, whose strangeness has challenged semioticians and philosophers, who have stated that it is an unclassifiable animal that escapes the preconceived categories of our minds—to such an extent that one of its first taxonomic names, provided by physician Johann Blumenbach, was "paradoxus."

The Australian aborigines thought that the platypus was born from a mythical cross between a duck and rat. The mistake is to regard it as a chimerical mixture, as if, in retrospect, a creator with a sense of humor had been playing games fusing together parts of a bird, reptile, and mammal. In reality, as Darwin had already perceived, the platypus is not an a posteriori construct but rather an a priori one in the sense that its monotreme ancestors split early on (about 160 million years ago) from the branch that later gave rise to marsupials and placentals. So the phylogenetic lineage goes back a long way, although it is neither a living fossil nor a missing link between mammals and reptiles.

For this reason, its genome, which was sequenced in May 2008 and was featured on the cover of *Nature*, contains a cluster of highly conserved genes that are found with similar functions in reptiles and birds (linked to egg formation and development, the production of venom, and sex determination through ten chromosomes). Electrolocation, on the other hand, is thought to have evolved more recently through the functional co-optation of genes for the sense of smell. In short, the platypus is an imperfect and special being with an unusual biology, strange but modern in its own way. It is the only remaining form of its family and genus, and it lives in eastern Australia and Tasmania, suffering, like all of us, from the effects of global warming and pollution. For the time being, however, it is not doing too badly compared to other endemics (forms of life, animals or plants, that live only in one specific place, such as an island, mountain, or forest area, and for this reason are fragile and vulnerable).

We can now enunciate a third law of imperfection, adopting Darwin's words in the sixth chapter of *The Origin of Species*. *Natural selection is not an agent that perfects and optimizes organisms in every part of their being. It cannot do so because it works in contingent circumstances and thus is always relative to a changing context, and above all it is conditioned by historical, physical, structural, and developmental constraints*:

> Natural selection tends only to make each organic being as perfect as, or slightly more perfect than the other inhabitants of the same country with which it comes into competition. And we see that this is the standard of perfection attained under nature. . . . Natural selection will not produce absolute perfection, nor do we always meet, as far as we can judge, with this high standard under nature. (Darwin 1872, 163)

Not even the human eye with all of its inimitable devices is perfect, Darwin continues, quoting physicist Hermann von Helmholtz. There is a long list of malfunctions: it forces the brain to make continuous adjustments and integrations, it has a retina that turns in on itself (the octopus's eye is much better, not to mention the four eyes of the jumping spider, or what eagles and hawks can see), it has only three color receptors (the compound eye of the mantis shrimp has from eight to twelve), it has a narrow field of vision, and it suffers from chromatic aberration, blind spots, and other unpleasant defects that all spectacle wearers well know.

Not to mention, Darwin insists, the imperfection of the bee's sting, which, used in self-defense, leads to the death of the insect itself. Or the thousands of poor, useless drones that are slaughtered. Or the waste of pollen from fir trees. It makes no sense. And yet these imperfections work. The sensitivity of our eyes is nonetheless incredible, and wouldn't we love to imitate their performance in cameras? When we admire the beauty of ichthyosaurs in natural history museums, what are we really appreciating if not the effectiveness and even aesthetics of opportunistic imperfection? We are observing a fish lizard that has done rather well for itself following its unusual choice to return to the seas. It shows us an entirely original mixture of elements induced by its adaptation to a marine ecology (i.e., its naturally imperfect "evolutionary convergence" with fish) and characters inherited from its direct descent from terrestrial reptiles. As an aspiring fish, or returning one, this is not optimal by any means. Some mammals did the same thing at the same time, evolving into cetaceans. It is not the ne plus ultra in perfection, but it works. Okay, the ichthyosaurs are now extinct, but they happily sailed the seas from the Triassic to the Cretaceous for 160 million years. With our meager 200,000 years of life, we are not in a position to make any judgments.

So it is in no way necessary and inevitable that every part of every organism carries out a specific optimal function. It could also be a consequence of other factors, such as structural or developmental constraints, or a past relic that refuses to leave. It is not necessary, then, to look for a functional explanation for fingerprints, for the red color of our blood, for the pink color of flamingos, or for why all terrestrial vertebrates have inherited five-toed limbs rather than six- or eight-toed ones. Perfection and elegance are not the criteria of nature. The important thing is that a structure works.

The omnipresence of rudimentary and imperfect organs in nature does not therefore pose a problem for the Darwinian evolutionary position. On the contrary, it provides confirmation of the universal common descent of all living beings. This also includes humans, with their useless earlobes, their tedious wisdom teeth, their protruding chin, their excessive nose considering what it is used for, their delicate and vulnerable skin, their vermiform intestinal appendage, their spinal curves, and their vas deferens, which carries sperm from the testicles to the penis not directly and by the shortest route but instead after going by a useless and lengthy route via the ureter, their imperfect nerve attachments on the spinal column, their coccyx, the remains of their ancestral quadrupedal gait, and the corresponding ills and pains, backache, sciatica, flat feet, scoliosis, and hernias. This is why Darwin's *The Origin of Species* promptly mentions the useless characteristics of the human body.

THE PROBLEM OF PERFECT ORGANS

There was another and equally important reason why Darwin struggled with perfection all of his life. His opponents pointed out that natural selection could not explain the origin of particularly complex and therefore perfect organs. Selective processes work only gradually, accumulating small variations from generation to generation. Furthermore, each evolutionary step of a specific structure must be functional and useful for its owners, who otherwise would not survive and reproduce. Without these two stipulations, Darwinian evolution does not work. Darwin's opponents therefore mischievously insisted on asking: How can natural selection explain the gradual onset and early stages of highly complex and perfect structures such as an eye? A rough draft of an eye cannot see. A rough draft of a lung cannot breathe. And a rough draft of a wing cannot make you fly, and a 5 percent camouflage cannot hide a prey from its predator. And so?

Many other scientists after Darwin had the same doubts: natural selection seems unable to account for the evolution of the later stages of particularly elaborate structures, where many parts must interact with each other and the lack of one component risks the failure of any possible adaptive advantage. As he so often did, Darwin took this criticism seriously, admitting that it was a real difficulty for his theory. He agonized over it for years, and eventually added more than an entire chapter to the sixth and final edition of his famous volume in 1872 in order to respond to this specific objection regarding perfection. He could not renounce either the continuity and gradualness of evolution (his enemies insisted on his admitting that the eye had evolved in one go, as if by some miraculous internal force), or functionality of the later stages (his enemies wanted him to accept the idea that a final cause, an intelligent plan, had been in progress right from the beginning).

So he decided to answer in two ways—mild but firm, as was his style. First of all, he explained that it is important to consider that adaptation is a dynamic of change, a current of transformation, and not a completed *optimum*. The function of the first eye, naturally, was not to see as we do today but rather to barely identify a source of light (and so distinguish between high and low sea levels, for example) or even glimpse a potentially dangerous silhouette, begin to see contrasts of light and dark, then later discriminate sharper outlines, and finally see three-dimensional objects, sharp colors, and perspectives. From gradation to gradation, driven by the strong selective pressure of orientation in space (if you have predators around, it is better to spot them early), the eyes evolved at least thirty times in parallel in different lines of animals. But they never evolved twice in the same way, never reaching the same result twice (such as lensed eyes, eyes with multiple tubes, pinhole eyes, compound eyes, etc.). Through the gradual accumulation of small yet useful hereditary variations, they transformed from rudimentary and imperfect eyes into increasingly complex and apparently perfect eyes. Not perfect in absolute terms, but comparatively more perfect.

The first Darwinian hypothesis, which has now been widely confirmed by countless case studies in nature, was thus based on gradual and relative improvement. Okay, we all agree that only 5 percent camouflage is not enough to hide a prey from its predator, but if the owners of that small variation have a better chance of coping than their companions that have no camouflage at all, this is more than sufficient to trigger a process of change toward increasingly effective and even spectacular camouflages (try looking for a stick or leaf insect hiding in a bush). This process is the root cause of many of the most amazing adaptations we can see in nature.

Darwin, however, also had a second answer because he considered the first to be insufficient. We can use this to draw up a fourth law of imperfection. According to the founder of evolutionary theory, the relationships between structures and functions in nature are generally redundant. A single function can be performed by several organs, which means that, when necessary, one of them can be co-opted for new uses with no adverse effects on the overall health of the organism. Conversely, a single organ can perform several possible functions, some of which might be already operational, while others are only potentially operational, but ready to be recruited when required.

This opens up the possibility of a second mechanism: parts of the organism that had previously been selected for a certain ancestral function (e.g., the ossicles required in fish to support the gill arch) can "readapt" to perform new functions (e.g., supporting the jaw and making chewing possible in the first terrestrial tetrapods). Naturally, selective processes cannot start from zero, as this would not be economical. Where possible, you can use existing material because a small, imperfect but immediate benefit is better than a vague and murky future imperfection. Moreover, it is difficult to discard existing material because an organism must keep living while waiting for a replacement. In this way, from generation to generation, selection converts structures from one function (in the previous case, breathing in water) to another (chewing). This mechanism then repeats itself; for instance, the three small bones of the middle ear in mammals derive from those that suspended the upper jaw from the skull in other vertebrates. So there are in this case three functions, all distinct from each other: supporting the gills, attaching the jaw to the skull, and facilitating sound transmission. In 2011, paleontologists found a transitional form in China in the form of a mammal that lived 125 million years ago, thereby illustrating this process of functional readjustment in action.

It occurs in plants, too. According to Darwin, the torsion capacity of climbing plants, which is also involved in their ability to coil and their sensitivity to contact (to different degrees and in various combinations from plant to plant), may have evolved by using the rotational movement of young stems for climbing purposes, which in itself is the result of a physical constraint and has no utility. For Darwin, proof of the functional co-option hypothesis lies in the relationships with other species and geographic distribution. In short, the great naturalist imagined the existence of a mechanism (now known as exaptation—i.e., reutilization from an already-existing form) that has proved to be an extremely widespread and powerful phenomenon in nature (Gould 2002; Gould and Vrba 1982; Vrba and Gould 1986).

Many crucial innovations in evolution—on the molecular, morphological, and behavioral levels—are exaptations. It was discovered that many genes were once linked to functions that were different from the ones they are involved in today. The toes on the limbs were already present in fish that had lobed fins and dragged themselves along shallow muddy beds long before they became a crucial adaptation for walking on land. Fins and feathers evolved in theropod dinosaurs to perform functions related to thermoregulation, sexual selection, and running balance, before being profitably reused for gliding and then active flight. As a result, many birds today continue to use their feathers for hovering, body temperature regulation, exhibition, and courtship. So we go from thermodynamics to aerodynamics gradually, but you can also go back on your footsteps; ostriches and the rare New Zealand kakapo parrots no longer use their wings to fly, but reuse them to balance their running, display their strength, court, and shade their young. Therefore 5 percent of a wing was not a wing, full stop. And wings did not evolve "for" flying.

This has several interesting consequences. First of all, when we study a structure, even a particularly perfect one today, we should not—with hindsight—immediately presume that its current function coincides with its historical origin. Perhaps it was once used for something completely different and then was co-opted. Second, we must recognize that if this is how evolution works, then any organ that we see performing a certain function at any given moment has the potential to perform other functions as well. Natural history is always full of possibilities.

Moreover, if the organ has not been gradually and selectively adjusted to suit its current function but rather is the result of ingenious evolutionary tinkering, then probably its structure is not perfect, in much the same way that any form of rehash is almost never perfect. But it works, and possibly quite well. Metaphorically, evolution works like a skilled craftsperson who makes do with what they have at the time rather than like an engineer or architect who has drawn up plans on paper beforehand. And so here is our fourth law of imperfection: *the reuse of already-existing structures means that suboptimal—that is, imperfect—structures are frequent in nature.*

POSSIBILITIES OUTNUMBER REALITIES

Darwin frequently stated that the saying "vox populi vox Dei" (the voice of the people is the voice of God) does not apply to science, which is often counterintuitive—above all, when it denies us such comforting concepts as perfection. We can observe the refined biomechanics of an eye and immediately the telescope can come to mind, given that it is a designed artifact (not perfect, however, but perfectible). This association comes naturally to us, but it is wrong. They are paradoxes of evolution, which lead us astray in our understanding.

Imperfection seems popular (nobody is perfect, and on and on with all the clichés), but in reality it is annoyingly counterintuitive. Yet it does in some way reveal how things went. Take, for example, the giant panda, the cuddly emblem of endangered or critically endangered species. Human greed, which has destroyed most of the ecosystems in which it lives, interacts negatively, like the end of the Ice Age did for the Irish elk, with its congenital imperfection. The panda is a bear—that is, a carnivorous mammal—but it eats bamboo from morning till night. How is this possible? Forcing your domestic cat to follow a vegan diet may not be such a good idea.

The panda is potentially omnivorous like other bears (in zoos, at times, it eats honey, eggs, fruits, and tubers), but in the wild it has been a near-perfect vegetarian for two million years, with bamboo shoots comprising 99 percent of its diet. Its digestive system, though, remains the apparatus of a carnivore, and it is only due to the special bacterial microfauna in its intestines that it is able to digest cellulose. Even its teeth are those of a bear. Like all herbivores, but without their specialized stomachs, it must eat constantly and in large quantities to gain enough energy because bamboo is obviously much less nutritious than an antelope steak. It must also lead a rather sedentary life, moving slowly and conserving its strength. How boring!

Much the same happens to the koalas; compulsive eucalyptus leaf eaters (but only of certain varieties, whose toxins they are able to break down), they have slowed down their biorhythms to the point where they sleep eighteen hours a day. The opening of their pouches faces downward—which is a distinctly bad idea for those that live high up in the trees and risk dropping their babies—but apparently their burrowing marsupial ancestor had one in this way and the koala has yet to change it. I know that you're thinking that the panda and koala were maybe asking for it, and that it is probably not all our fault that they're in danger of extinction (we *Homo sapiens* are great at passing on the blame). In fact, the panda and koala illustrate our fourth law of imperfection very well. They changed their eating habits and adapted as well as they could; they are peculiarities that work. They are just two of the countless examples that demonstrate the importance of being able to make do in evolution.

Certain anatomical details are even evidence of makeshift adjustments. It is difficult for the panda to grasp bamboo with a bear paw. Natural selection has therefore favored individuals with greater prehensility; over time, the panda has developed an opposable "sixth finger" starting from a small bone in its wrist, the radial sesamoid. This is not a genuine thumb *ex novo*, but rather an opportunistic reutilization. A structure created to perform certain functions is later co-opted to perform other, completely different ones, following changes in environmental circumstances and, in this case, diet. The panda is the result of evolutionary tinkering (Gould 1980, 1993).

But it does not finish there. The corresponding bone found in the foot of the panda, the sesamoid of the tibia, has got bigger too, developing symmetrically with the upper limbs, but with no particular utility and perhaps even with some discomfort for a plantigrade such as the panda. Imperfection thus also shows us the genetic correlations of development that transform our bodies into a coordinated unit. If an adaptive change takes place in one part of the system, we must expect side effects elsewhere.

Darwin was persuaded by this second answer to the problem of perfection to the point that he took it as a sort of general principle in the later years of his life. In Darwin's ([1862] 2011) wonderful book on orchids, he wrote that in the whole realm of nature almost every part of every living thing has probably served, with a few modifications, other purposes and functioned as part of the living machine of many and varied ancient forms. In this respect, his favorite word was "contrivances." The underlying concept here is the idea of improvisation. Nature does not make plans, it finds solutions. Darwin also discussed functional shifts in the evolution of marine animals with his friend Anton Dohrn, who founded the Naples Zoological Station in 1872.

We can appreciate the same concept by generalizing it. Remember the ideal sponge we mentioned earlier? Evolutionists have invented a fascinating concept that they call "morphospace." For a certain body shape (of a crustacean, insect, and so on) or trait, such as the shape of the shells of land snails, one can mathematically construct an imaginary space that contains all the combinations of possible shapes. In practice, this represents the overall space that evolution can explore, and can be reproduced on Cartesian axes using quantitative parameters. This ideal morphospace can then be compared to reality—that is, with the forms that the body plan or trait adopts and has adopted in all the species that actually exist and have existed. Through this technique, an interesting fact was thus discovered: evolution has almost always explored only a small subset of possible scenarios. The chosen solutions are usually limited to one or a small number of restricted areas of morphospace. Why?

The first answer that comes to mind is that those areas are ideal because they are where the optimal forms are concentrated—that is, the best adaptations or the peaks gradually reached by natural selection (Dennett 1995). In some cases this is true, but rarely. Species have sometimes coped well in regions that are in no way excellent, but simply acceptable. There are also entire regions of morphospace that would have provided a good chance of survival for those species that explored them, but no such exploration ever took place. We can assume then that no animal ever adopted those solutions because of preexisting physical, structural, or developmental constraints that precluded access. Or we must accept the idea that it is sheerly by coincidence that historical contingency has so far never led anyone there. The right genetic mutation, for example, may never have come up on the roulette wheel. In any case, imperfectly functioning potential alternatives have always existed and always will.

"There are more things in heaven and earth, Horatio, than are dreamt of in your philosophy," playwright William Shakespeare made Hamlet say. There are more things in heaven and earth than evolution has ever dreamed of. The possible is much greater than the real. And nature is greater than all the theories that we have come up with in our attempt to understand it.

4

THE IMPRINT OF USELESSNESS IN DNA

> "Well! My dear Pangloss," said Candide to him, "when you were hanged, cut up, beaten, and rowed to the galleys, did you still think that everything was going well?" "I still hold to my first opinion," replied Pangloss, "because, in short, I am a philosopher: it does not suit me to contradict myself, [philosopher Gottfried Wilhelm] Leibniz not being able to be wrong, and harmony being the most beautiful thing in the world, just like solid and ethereal matter."
>
> —Voltaire, *Candide, or Optimism*

In evolution, then, is nothing at all thrown away? It depends, as always, on unstable compromises. If the defect or excess becomes expensive for its owner, and is detrimental to survival and reproduction, natural selection will most probably wipe it out. Within a few generations, the variation will have disappeared from the population or will persist at a low frequency. Yet the opposite can also happen—that is, a trait that has become useless happily remains widespread in the population, totally ignoring selection, for a simple reason: because it would be too expensive to remove it. We might as well keep it if it does little harm. This tolerance toward imperfection is most often found where we would least expect it: in our DNA.

ATAVISTIC GENES AND THE CHICKENOSAURUS

As we saw in chapter 2, in all of DNA's amazing self-replicating ingenuity, one would instinctively consider its sanctum sanctorum to be the kingdom of perfect organizational efficiency. In reality, we have already said that a two-faced Janus lurks in the shadows, namely genetic mutation, which provokes constant swings between stability and variation. If we take a quick look inside, we discover that perfection is by no means at the top of its list of priorities.

First of all, there are many dormant genes in the genome of animals, and they sometimes erroneously reawaken in the development process—that is, they are activated when they should not be. This means we often, for instance, see the birth of whales equipped with rudimentary and strange hind legs, which are obviously completely useless. To take another example, at times snakes are hatched sporting little legs. In ancient times, horses born with three toes and hooves, a rare event, were highly prized. All are sports of nature.

These dormant genes are also called atavistic because they bring back traits that had previously been lost in the course of a species' evolution. They are the genetic counterpart of the vestigial traits we mentioned in the previous chapter. In a way, they are genes that turn back the time machine, or archaeological genes. Like all mammals, we have them too. In the human genome, as many as two thousand deactivated genes or pseudogenes have been identified—genes that were functional until they underwent mutations that switched them off (e.g., genes responsible for the internal production of vitamin C, which primates have learned to obtain through their diet, or genes responsible for olfactory receptors, which are largely deactivated because our sense of smell is no longer so important for us). So every now and then a child is born with a small tail, or more precisely, an atavistic, extremely elongated coccyx (which is nothing serious, as it can be easily removed).

We even have the archaeological remains of three genes that enabled our remote reptilian ancestors to produce egg yolks. In the early stages of development, our embryos form an anachronistic vestigial yolk sac that later disappears, and it is another reminder of a distant past still written into our DNA. What's more, we still have a genetic input that causes the growth of a thick downy layer in six-month-old fetuses. This serves no purpose, but is a legacy of the hair we have lost, and in fact it disappears a month before birth. But if these genes have no current function, or they generate useless embryonic structures that then disappear or occasionally remain at birth in the form of atavisms, what are they for? It is clear that evolution works by adding the new onto the old, or building the new on top of the old, and does not wipe the DNA slate clean of everything that is no longer needed. It frequently maintains the unnecessary, opting simply to avoid its expression by silencing it.

Genetic atavisms can also be exploited to do strange things, such as trying to resurrect dinosaurs. Prominent paleontologists such as Jack Horner are working with geneticists to regenerate some dinosaur traits through the genetic modification of chickens. The idea is simple, and it works because of the redundancies and imperfections found in DNA: birds are evolved and modified forms of dinosaurs, so they automatically have some atavistic, useless genes in their genome that were responsible for dinosaur traits and then switched off. All you need to do is to find them, reactivate them, and create your chickenosaurus.

It sounds very much like *Jurassic Park*, and in fact Horner was the film's scientific adviser. The first concrete results have reduced the initial skepticism of the scientific community. In

2009, biologist Matthew Harris at the University of Wisconsin found an atavistic gene responsible for teeth and produced the first toothed bird. Having hens with teeth is not such an absurd concept because they still have the dinosaur teeth genes in their genome, and all that is missing is a protein to reactivate them. If you supply hens with one, the teeth will return to the beak! In one lab in Chile, scientists managed to regenerate the fibula bone in a chicken leg, and in another lab, at Yale University, they modified the skull of a bird so that it resembled a dinosaur. Horner is now trying to give birds a long dinosaur tail again, but he has found this more difficult. Nevertheless, he says that his chickenosaurus will have become a reality by 2024.

In the meantime, thanks to these genetic investigations we are learning a great deal about how birds evolved from nonavian dinosaurs. Moreover, we are gaining interesting insights into some genetic diseases. At some point, we must decide what we should do with these dinosaur-birds, considering that the environment has changed completely since the Cretaceous period. Thanks to state-of-the-art gene editing, we might be able to resurrect mammoths, thylacines, and other extinct species. The danger is that they will become circus freaks, generating a lot of business and not much science. Or as Horner proposes, maybe the chickenosaurus will eventually become nothing more than another pet. We will breed them and let them run around in the garden, much to the delight of our children. Unless the story ends up like *Jurassic Park*.

THERE ARE MANY TYPES OF JUNK

But where does this tolerance of DNA toward extra genes come from? In 1998, the great molecular biologist and Nobel Prize winner Sydney Brenner introduced an intriguing principle, distinguishing in English between two concepts of *junk*. He pointed out that there are two types of rubbish in the world, and that most languages have two separate words to distinguish them. There is the rubbish we store, which we call "junk," and the rubbish we throw away, which we call "garbage." Brenner said that the excess DNA in our genome is junk, and it is there because it is harmless as well as useless, since the molecular processes that generate excess DNA outnumber the processes that eliminate it. If excess DNA became disadvantageous, it would be subject to selection in the same way that junk that takes up too much space, or is beginning to smell, is immediately converted into garbage.

Brenner is telling us that there are two types of excess in DNA. The first is equivalent to the dusty junk that messy persons accumulate in their garages, much to the scorn of their partners. When asked to explain this irrational attachment, messy people usually say that it costs nothing to keep it there, they are fond of it, and maybe one day some of it will come in handy. There are also atavistic genes in that jumble—old, once-useful tools that remain in the garage because you never know. Then there is the bulky or even smelly rubbish, usually organic, which has to be sorted and left at the garden gate awaiting waste collection.

Now, why does DNA, which has the delicate task of transmitting genetic information and is the driving force behind the evolution of life on earth, keep a lot of stuff in the garage like untidy people? Well, there are a number of reasons. First, it's because old and unneeded tools can have an unexpected utility. Uselessness is a source of innovation, even in DNA, because genes that have lost a certain function, or have several, can be recruited to carry out new ones. Let us assume for a moment that a gene is involved in a crucial aspect of an organism's physiology. Being so valuable and essential, natural selection will guard it carefully night and day. Metaphors aside, this means that it is highly likely that any mutation in that gene will produce negative effects and therefore be cleaned up over generations. Such a gene is said to be "conserved," or taken care of, by evolution. If, however, in the course of natural history that gene had by chance been duplicated over and over again, its redundant copies would be of absolutely no use. A large number of neutral mutations could thus accumulate on these copies—insignificant for survival given that those copies of the gene are nothing more than a surplus.

Now let us imagine that one of these copies, again through chance as well as after considerable trial and error, acquires a mutation that improves the performance of that gene through a slightly modified protein or makes it possible for that gene sequence to become useful in other important functions. At this point, we have an improved version of the gene in question, one that will be favored by selective processes and become successful. Basically, DNA invented backup copies long before computer scientists! The original copy of a useful gene continues to do its job, while mutations can work freely on the backup copies with no form of risk. As such, sooner or later, one of the copies could undergo a favorable mutation and will be co-opted to do something else.

Remember the concept of exaptation, or opportunistic recycling, that we came across in the previous chapter when talking about panda thumbs and the ingenious reuse of wings? Well, it also happens in DNA. In the course of evolution, a gene can be "refunctionalized," which is to say regulated differently, and co-opted for new uses. In 2016, a team of neurobiologists coordinated by Michael Greenberg at the Harvard Medical School published some experimental results in *Nature* that provide a fascinating interpretation of the genetic basis of human brain evolution (more on this in the following chapter). The researchers cultured human and mouse neurons, and then compared their responses to a type of stimulation that simulated increased neural activity. In this way, they could observe which genes in the two in vitro cultures were most affected by excited neural activity.

Many genes, especially those that respond to stimulation immediately, are the same in both species, as expected considering the genetic similarities between mice and us. There is one in particular, however, that is activated later and almost exclusively affects the neocortex, which has a crucial difference: it is the osteocrin gene, which is known to play an essential role in bone growth and muscle function in vertebrates. This gene is not expressed in the mouse brain or induced through stimulation in mouse neural cultures. It carries out its work in the bone and muscle but shows no activity in the rodent brain. In contrast, in human neural cultures, it is strongly expressed in the neocortex and above all in mature neurons in the developing cortex.

What is a gene connected to bone and muscle activity doing in the middle of the part of the brain involved in the most complex cognitive functions, including thought and language? The researchers continued their studies and discovered the genetic changes that lead to the activation of osteocrin in the brains of those primates that are most similar to us, but not in those of other mammals. Through a few mutations in the promoter regions, the gene has switched function in the course of evolution. As in tinkering, it has been reused or recycled from the bones to regulate the shape of dendrites, promote axon elongation, and facilitate other structural changes that neurons undergo during the learning period (Ataman et al. 2016). So the evolution of our dense neural network also depended on genetic exaptation. After all, isn't it cheaper and easier to transform what is already there than to build something completely new?

To make these processes possible, DNA must put up with a certain level of redundancy. Were every gene to be crystallized in the coding of one and only one protein structure, linked to one and only one function, there would be no room for maneuver. What is more, apart from the DNA itself, genetic junk is useful to the scientists who study the genome. As already mentioned, since natural selection has no interest in them, the extra parts of DNA tend to accumulate a large number of mutations with a certain regularity that will be neutral regarding the survival of the organism, and therefore tolerable. Scientists can then use pseudogenes and their harmless mutations as a "molecular clock" because if we know that a certain gene was deactivated in the common ancestor of two species, and then count how many different mutations have accumulated in them, we will be able to calculate the time when the two species split.

In practice, neutral mutations on these nonfunctional genome traits act as a ticking molecular clock that can be used to measure time and kinship between species. Many of these atavistic genes were silenced in the early stages of a species' evolution, following changes in environmental conditions, and so similar genes can probably also be found in related species in which they may still be active. Thus archaeological genes are a source of valuable evidence in the reconstruction of the genealogy of species—the tree of life.

FROM JUNK DNA TO JUNGLE DNA

The metaphors we use when referring to DNA have one defect: they are all two-dimensional. Texts, the information, the book of life, the alphabet, the code, and the software. They are centered around one principal aspect of genetic transmission, its linearity of sequential reading (and now rewriting), but they miss the fact that the genome is itself an evolving system, a three-dimensional material system that has transformed itself through time, obeying the same *laws of imperfection* that we have seen applied to organisms and species. Furthermore, it is not an isolated machine but rather is part of a dense network of relationships and regulations at genetic, epigenetic, cellular, tissue, organ, organism, and external environmental levels, similar to a matryoshka doll but in constant turmoil.

If we take a look inside, the genome as we know it today looks like a jigsaw puzzle, but a puzzle in which each piece is connected on the underside to many other pieces via fine threads that we mostly cannot see. Moreover, a substantial proportion of these pieces have no fragmentary picture painted on the top that can be connected to any of the others. They are white, enigmatic, and apparently useless. In 1972, the Japanese geneticist Susumu Ohno coined the term "junk DNA," in the first sense given later by Brenner (though molecular biologist Francis Crick had already imagined something similar), to define those large portions of genetic heritage that seemed indifferent to the action of natural selection (neutral). Junk DNA was defined by Ohno as any segment of the genome that has no immediate use, but that might acquire some function in the future—obviously without any form of foresight, but through, for example, gene duplication mechanisms such as the backup copy technique we saw earlier (and that was in fact theorized for the first time by Ohno in 1970).

The problem is that there is a lot of junk DNA. Too much. Embarrassingly, the percentages are comparable to those of dark matter and dark energy for physicists who are studying the expansion of the universe. Junk DNA immediately appeared to many as the true statistical dominator of the genome, a remnant of nature's failed experiments that remained in the evolutionary system because the molecular processes that generate the extra DNA, as Brenner said, are evidently more powerful than those that eliminate it. As long as it does not disturb anything, natural selection leaves it in peace. But the fact that it tolerates so much of it is impressive and a little perplexing.

In the meantime, the first data regarding the complete genome sequencing of various organisms began to emerge. The surprise grew further in the second half of the 1990s and early years of the new century, when geneticists realized that they had been totally wrong (by an order of magnitude!) in their predictions concerning the number of human genes. The Human Genome Project revealed that there were not 250,000, and then 120,000, as had been assumed but instead less than 25,000—rather few. The original metaphor was wrong: the genome is not a sack of balls, each connected to an external trait of the organisms, but rather a densely interconnected network. It is not so important how many nodes there are in a network as it is how they are linked together. On average, one gene can produce four different proteins in different tissues. The number of possible combinations of the 25,000 genes is astronomical. Remember the strange circularity of protein synthesis? Well, in the genome, the controller is in turn controlled and regulated by its own products. This means that there is no direct proportion between the complexity of an organism and the number of genes it has.

Junk DNA triumphed when it was discovered that only a small percentage of the hereditary material is made up of genes coding for proteins (much less than 10 percent), with the rest being surplus. The human genome is indeed redundant, full of waste material and background noise. As one geneticist put it, the tiny fragments of DNA that code for proteins, and therefore have a known or predictable function, float like makeshift rafts in a vast, meaningless genetic ocean.

But another twist in the story was soon to follow. According to the international consortium that has spent the last ten years compiling ENCODE: Encyclopedia of DNA Elements, based on an analysis of transcripts (i.e., the products of genes), it is true that maybe less than 2 percent of the human genome is made up of protein-coding genes, with a larger proportion (between 9 and 18 percent) linked to regulatory functions. This confirms that it is their relationships and regulations that are important as opposed to the number of nodes in the network. Thus there are treasures hidden in junk DNA, which in particular include the sequences that transcribe for the many forms of noncoding RNA involved in the intricate web of gene regulation, a dense network of connections that we know little about. And in the midst of this tangle of gene products, we can find the causes of many diseases, including the dynamics of tumor transformation.

After the first results were published in 2007, the hundreds of scientists involved in the ENCODE project went on with their work, eventually reaching an even more radical conclusion, which was published in September 2012 amid much publicity in *Nature*. The researchers stated that as much as 80 percent of the genome is transcribed in RNA and hence they suggested that it could be biochemically functional. We don't know what it is for, but it has a function. The message was clear: ignore what was previously said; the apparent uselessness was due to our ignorance regarding the complexity of the genetic code. This is typical of science; thanks to new studies, we realize how much we did not know. Junk DNA is a misleading concept, which should be abandoned after a distinguished forty-year career.

"Junk DNA Is Dead" was an emphatic headline in all the newspapers. Does this mean the genome once more resembles a more efficient system? Are there patterns in the genome, or perhaps even a hidden language, that we had not seen? For what mysterious reason should the rules of evolution for regulatory elements (which change at a faster rate) be different from those that apply to protein-coding elements? ENCODE's results did not go down well with many other biologists, who in an article that appeared in the months that followed, attacked the theoretical basis of the project. Dan Graur, a prominent molecular biologist at Houston University, wrote that their statistics were really weak (Graur et al. 2013).

Having a biological activity (to be transcribed) does not mean tout court having a function, according to the dissenters. The ENCODE team almost exclusively used pluripotent and tumor cells, which are permissive and specific environments when it comes to transcription. This might have influenced the results. The estimates are also imprecise, and the whole paper fails to put forward any plausible hypothesis regarding how those noncoding but transcribed parts could have been preserved in evolution without natural selection. We do not know whether the extensive and intense transcription activity is an irrelevant background process (vestigial or random), or whether all the RNA produced performs a function we have yet to discover, but the latter hypothesis is highly unlikely considering that pseudogenes have transcripts too. The raw data of bioinformaticians must be translated into knowledge and placed in a correct interpretative context—an evolutionary context.

Graur and colleagues' reply ended venomously. By insisting on looking for functions where none exist, we end up moving from functions to the idea of a purpose, of a project:

> We urge biologists not to be afraid of junk DNA. The only people that should be afraid are those claiming that natural processes are insufficient to explain life and that evolutionary theory should be supplemented or supplanted by an intelligent designer. ENCODE's take-home message that everything has a function implies purpose,

> and purpose is the only thing that evolution cannot provide. (Graur et al. 2013, 33)

Whatever one thinks about junk DNA, the fact is that we are weighed down by generative ignorance and all we have yet to understand, thereby producing new research questions. The world of gene transcripts, in particular, is revealing an extraordinary wealth of information, which is difficult to interpret. As Ewan Birney, the coordinator of ENCODE at Cambridge, eventually admitted, the genome resembles a jungle, a dense forest, a wall of elements through which you have to fight your way. You try to beat your way through to arrive in a certain place and you're not really sure where you are. It is highly likely you will feel lost in there. From junk DNA to jungle DNA, we have a baroque genome.

THE LAW OF THE ONION

All things considered, the news concerning the death of junk DNA might then be somewhat exaggerated or at least premature. Further data confirm this notion. How can we explain, for example, the fact that the genome of an onion is five times larger than that of a human being? It would be difficult to argue that the onion is five times more complex than we are. It is far simpler to point out that evolution in plants often involves speciation, namely the birth of new species through hybridization, and subsequent fusion of the genomes of two or more parental species. That means that the DNA becomes much larger and heavier, as can be seen in soft wheat, which now has three times as many chromosomes because it was the product of the reckless fusion of herbaceous plants following the experiments of the first farmers in the Fertile Crescent. Soft wheat, a monstrous genetic giant, has become our daily bread.

As geneticist Timothy Ryan Gregory has argued, the onion test is a simple reality check for all of those who think they can assign a function to every nucleotide present in the human genome. The size of the genome can vary greatly in different organisms of similar complexity. Evolution always takes place at several interconnected levels. As in the case of the gut microbiota, we can imagine that the jungle of DNA is also an authentic ecosystem with its own rules and mechanisms that do not depend solely on the effects on the organism (Gregory, Elliott, and Linquist 2016). Some redundant sequences are repeated so many times that they actually occupy a large part of a chromosome, as if they were a sort of invasive species or virus. It is evident that they follow their own "selfish" logic, which only makes sense at the level of their genetic ecosystem and not at the higher level of the organism. This too represents an unstable compromise between processes acting at different levels.

So DNA is information and code, but it is also three-dimensional matter. The genome is a concrete evolutionary system, and as such it is imperfect. It would be considered a failure as a product of engineering, proving that it is a product of Darwinian evolution. If we forget this aspect, we run the risk of falling into the apophenia trap: the human tendency to see meaningful figures and patterns in a sea of data that is in fact random, or to find functions at all costs even where none exists. Some philosophers, such as Daniel Dennett, think that evolution is an example of ingenuity at the highest level, which may well be true, but not always. There is a risk of applying our anthropomorphic engineering ideals to nature. As philosopher of biology Peter Godfrey-Smith has noted, to consider natural selection and the creation of a complex design as the fulcrum of evolutionary theory is to take the problem created by Paley and the theologians and to simply replace their solution with a naturalistic alternative. Continuing to focus only on the problem of the complex design means that we are adopting naturalism, but still remaining within the intellectual framework of natural theology. In a rain forest at dawn, in the ingenious transformism of an octopus with its chromatophores, in the architecture of DNA, there is no hidden project that is the result of research and development. All we have is history, with all of its twists and turns.

DNA is exuberant, and this is no coincidence. After organizing the protein-coding and regulatory regions, it produces and maintains much more material than it needs within itself. For the most part, the growing genome is made up of multiply repeated duplications, frequently repeated short sequences, pseudogenes or defunct genes, noncoding regions that space out coding regions, and disorderly jumping elements (active or inactive) that wreak havoc in the DNA and cause mutations. The latter, the "transposons," are the most selfish elements of the genome—those that think only of multiplying and spreading as much as they can—and they represent another mechanism that generates disorder, instability, and excess, but also, within certain limits, variability. There are several million of them in the human genome. They are often harmful when in excess and must be kept under control, but sometimes they have been co-opted in the course of evolution for important regulatory functions.

Our DNA also contains surplus material because it incorporates pieces of genetic material from outside. For example, retroviruses, which in the past infected sex cells by inserting a copy of their own genome, were then passed on to descendants and defused through other mutations. It has been estimated that almost a third of human genetic material is of exogenous origin. Thousands of these harmless genetic stowaways remain within our genome as molecular fossils—a further sign of our evolutionary past and of the ancient infections that we have survived. It is also possible that, in the course of evolution, some of these sequences have been reused for essential functions, such as the formation of the placenta in mammals (placentates).

Rather surprisingly, in recent years we have discovered that a further contribution to our genetic richness came from two other human species, namely Neanderthals and Denisovans. This is because our ancestors from Africa occasionally mated with them in the Middle East, Europe, and Central Asia in the period between about a hundred thousand and forty thousand years ago. Evidently, the genetic barrier had not yet definitively closed between us and them, and thus we were able to interbreed and give birth to hybrid offspring that were not sterile but instead able to have offspring of their own. The result of this activity is that this hybridization introduced some Neanderthal and Denisovanian genetic sequences into the genome of some non-African modern *Homo sapiens* populations, traces of which then became diluted and fragmented over time but still remain today. It is not yet clear how much of the DNA of other humans that ended up in our own DNA has been beneficial to us, as some evidence suggests, or slightly harmful, or inert.

Surplus genetic material could therefore fall into three evolutionary categories, which can be linked to the various mechanisms that produce extra DNA: the genetic egoism of some sequences, which are able to make copies of themselves; useless, but only within certain limits, as tolerable side effects of other functionally relevant processes; and potential reserves of variability and evolvability (i.e., the ability to evolve), which can generate positive (unforeseen) effects in the event of functional co-option. Naturally, assuming that evolution by natural selection cannot be influenced by future benefits, the third mechanism (being a source of variation for possible evolutionary novelties) is not the function lying at the origin of redundant DNA but rather a consequence of it.

The Nobel Prize–winning French physiologist and geneticist François Jacob described the evolution of the genome as a continuous form of tinkering, using parts that are recycled and reutilized for new functions. DNA is astonishingly universal, the basic biochemical unit of all forms of life, always consisting of the same polymers, nucleic acids, and proteins, the same four bases and twenty amino acids, whether in bacteria or whales, viruses or elephants, and it is this universality that suggests that its evolution must have taken place less by the addition of new elements and more through the continual reuse of existing material—that is, by the continual remixing of a limited number of elements (Jacob 1999).

Natural history is characterized by wonderful diversity, "but once life had started in the form of some primitive self-reproducing organism, further evolution had to proceed mainly through alterations of already existing compounds. New functions developed as new proteins appeared. But these were merely variations on previous themes" (Jacob 1977; all quotes from this source translated by Michael Gerard Kenyon). Variations on known themes and combinations of modules, as in the case of the few developmental genes that were already active at the time of the Cambrian explosion, have since then regulated the formation of the body structures of all animals, from zebra fish to *Homo sapiens*.

Evolution, then, is just *tinkering* with what already exists, for a different use and regulation of the same structural information, like a "giant Meccano" according to Jacob (1977). It works, above all, on previous evolutionary inertia. This is why animals that are morphologically and behaviorally different, such as chimpanzees and us, can have a genetic makeup that is actually quite similar (more than 98 percent identical). As Jacob (1977) states,

> Small changes modifying the distribution in time and space of the same structures are sufficient to affect deeply the form, the functioning and the behaviour of the final product—the adult animal. It is always a matter of using the same elements, of adjusting them, of altering here or there, of arranging various combinations to produce new objects of increasing complexity. It is always a matter of tinkering.

So in the name of tinkering and the redundant elements that permit it, we have a possible fifth law of imperfection: *excess, if it can be tolerated, is a source of change because evolution involves the transformation of the possible*. Let us call it the law of the onion, in honor of the humble and versatile bulbous plant that, when it comes to genes, outclasses us by a long way. Let us finish with the proclamation of the law by Jacob (1977) himself, made during a memorable lecture in Berkeley, California, because no one could put it any better:

> Often, without any well-defined long-term project, the tinkerer gives his materials unexpected functions to produce a new object. From an old bicycle wheel, he makes a roulette, from a broken chair the cabinet of a radio. Similarly, evolution makes a wing from a leg, or a part of an ear from a piece of jaw. Naturally, this takes a long time. Evolution behaves like a tinkerer who, eons upon eons, would slowly modify his work, unceasingly retouching it, cutting here, lengthening there, seizing the opportunities to adapt it progressively to its new use.

5

A PROVERBIAL JUMBLE: THE HUMAN BRAIN

> This speech gave rise to new thoughts, and Martin in particular concluded that man was born to live in the convulsions of restlessness or the lethargy of boredom. Candide did not agree with this, but did not affirm anything. Pangloss admitted that he had always suffered terribly, but having once claimed that everything was fine, he continued to maintain it, even though he did not believe it.
>
> —Voltaire, *Candide, or Optimism*

Let us suppose for a moment that you are a poor fish forced by circumstances to live in increasingly shallow and stagnant water that is lacking in oxygen. You have to breathe. Most of your fellows have not made it and are dying a slow death in the next pool. In your offspring and those of some of your fellow creatures, however, there is a blessed mutation that affects the wall of the esophagus, making it possible for them to breathe in air and absorb a little more oxygen. This slight advantage gradually spreads through your group. As the generations pass by, the breathing wall becomes a vascularized diverticulum, positioned above the digestive system, and later a bladder that rhythmically fills with air, and finally a true lung, of which there are countless anatomical variants. You have been the stars, together with a myriad of your peers, of an evolutionary transition. And, as always, necessity was the mother of invention. Looking at the process as a whole, as paleontologists are doing millions of years later, and making an adjustment for all the lost battles and all the failed attempts, we can say that an outgrowth on the esophagus has become a lung, demonstrating the power of tinkering. It's not perfect but it also works. Evolution is the transformation of the possible.

Could this have happened to the brain as well? Is this another evolutionary masterpiece that should be classified as an "imperfection"?

A BELATED BRAIN

The great Italian neurologist Rita Levi-Montalcini claimed that the brain of an insect was the perfect brain: so primitive, as small as a speck of dust, and yet it has coped stably with environmental problems for six hundred million years. This is precisely the opposite of the wonderful but imperfect brain of *Homo sapiens*, the outcome of a game of mutations that is unstable and therefore more creative and ambivalent (Levi-Montalcini 1987). Levi-Montalcini hypothesized that the vertebrate brain has always been an imperfect machine, and thus subject to continuous and inconclusive remodeling by variations and selective pressures. Invertebrates, on the other hand, were fine right from the beginning. It follows, then, that no insect will ever give birth to an Adolf Hitler or Albert Einstein.

There is nothing more graceless and fragile, not to mention ambivalent and unpredictable, than our brain. This was also Jacob's (1977) view: "The human brain was formed through the accumulation of new structures on top of older ones. A neocortex was added to the old olfactory brain of the lower mammals, which quickly, perhaps too quickly, took over the main role in the evolutionary path that led to man." That's right, *too quickly*! We will come back to this point in the following chapters. For now, let us just consider the strange anatomy and functioning of our brain, and not only ours, and evaluate them on a strictly evolutionary level. What is the cause of all this neuroimperfection?

First of all, we have a selective and hence partial brain. We only perceive a small part of the electromagnetic spectrum. As primates, we heavily invest in our vision and sense of touch, a little less in our hearing, and not much in our sense of smell. Even the information from the outside world that reaches us through our eyes and hands is limited and fragmentary, however—so much so that our brain actively filters, organizes, and analyzes it internally. The resulting interpretation is also partial and fallacious because it is in turn conditioned by past experience as well as by evolution in ecological niches that no longer exist. The result is endless numbers of optical illusions and errors in perception. We have materially limited sensations. So it is inevitable that we have unconscious and often misleading prejudices. We see the world through our own little slit window, just as all animals see it in their own way. But how did we arrive at our particular and imperfect perceptual and cognitive universe?

It is difficult to say. The problem is that brains do not fossilize, so we have no idea about the internal organization of our ancestors' brains in Africa. We can compare our brains today with those of chimpanzees, but we are talking about two cousins separated by six million years of neural evolution (and we know little about their evolutionary journeys due to the limited numbers of fossils) (Gee 2013). Or we can analyze the impressions left by ancient brains on the inner wall of the skull—the gibbosities, protuberances, and encumbrances that housed the meninges, sulci, gyri, blood vessels, and circumvolutions of the outermost cortex. And then we can speculate on the intelligence of a species based on indirect evidence concerning its technological abilities, adaptation to the environment, and sociality (but be careful because our social cohesion is nothing when compared to bees, termites, and ants, whose pinhead-sized brain, in the words of Darwin in *The Descent of Man* [(1871) 1981], is a marvelous atom of matter in the world). So we have to work our way through the clues, but it's better than nothing. So let us see what we currently know about the natural history of the human brain.

The first oddity regards the timing. It seems that our ancestors were in no rush to carry around a large brain. The intricate history of our subfamily began in Africa around six million years ago, when hominins separated from the common ancestor with chimpanzees. On the geological timescale six million years is not all that long, but it is long enough to accumulate a fair amount of changes in behavior, morphology, and posture. For decades, scientists pondered on which was the crucial innovation introduced by our early ancestors, and their attention centered on our wondrous head. It was presumed that the growth of the brain and the increasing intelligence that came with it had been the great vector of the human adventure—a wrong presumption, it turned out, because the scientists were looking in the wrong place.

Yes, wrong indeed, because the more than twenty hominin species that lived at the same time as us did not bother with a large brain for more than two-thirds of their history. Two-thirds of human evolution was spent in the presence of other species whose brains, apart from the differences between males and females, had a volume of about a third of ours. These included *Ardipithecus*, australopiths like Lucy and *sediba*, and the robust *Paranthropus*—the whole array of our possible ancestor species, from six to two million years ago, each bipedal in its own way, but all invariably with a chimpanzee-size brain. What's the point? If the brain is such an important organ for our survival and success as social primates, why did it start evolving so late?

The secret lay elsewhere, at the other end of the human body, in the feet (with painful compromises here as well, as we will see in the next chapter). Around two million years ago at the earliest, and only with the appearance of the genus *Homo*, did the growth of the brain finally get significantly started. Better late than never, but at that point at least, the journey led to triumph, up to the apotheosis of the wise human. Or so it was written in the textbooks, but even this was not true. Until recently, there was unanimous consensus that the growth of the brain of the genus *Homo* had been a trend—that is, a progressive and apparently unstoppable trend. But this is false.

TWO SMALL BRAINS DEBUNK A THEORY

First of all, it was not a solitary story. In Africa 2.5 million years ago, at the beginning of the phase of climatic instability known as the Pleistocene epoch, many different species, belonging to no less than three separate genera, cohabited in a territory that stretched from Eritrea to the tip of South Africa. The forest habitat shrank, the transitional grasslands (at the edge of the forest) favored by the australopithecines disappeared, and it is possible that the first hominin forms emerged that were adapted almost exclusively to open spaces. Over time, two alternative "models" of adaptation to the same fragmented environment emerged: the "*Homo*-like" model, with a moderately omnivorous diet, stone tools, and a gradually larger brain; and the "*Paranthropus*-like" model, with a specialized coriaceous vegetarian diet and huge, flat chewing teeth.

In the famous but elusive *Homo habilis*, the lower limbs lengthen, the bones become lighter, the face flattens, the cranial volume becomes one and a half times bigger (in the most recent specimens) than the australopithecines, the palate becomes more rounded, and the teeth highlight a mixed diet. Moreover, *Homo habilis* sites have an abundance of sharp stone chips, obtained by striking a cobble with a stone hammer. So, does the brain grow with technological skills? This is doubtful as there is nothing that excludes the possibility that in earlier times, other species might have used wooden or bone tools that have not been recorded in the fossil record. Furthermore, in 2015, a complex and differentiated lithic industry was discovered on the western shore of Lake Turkana in Kenya, and it has been dated to around 3.3 million years ago, or 700,000 years before the oldest stone tools found so far that have been attributed to the genus *Homo*. Whoever made them probably didn't have a large brain.

It is true that together with the subsequent growth of the brain, in the sites of the genus *Homo* and only in those sites, we can observe all the signs of the first complex behavioral system consisting of knowledge regarding the physical properties of the materials to be used, manual skills in detecting the fracture points in the stone, the ability to coordinate sensory-motor skills to avoid injury while working, and skills in passing on knowledge to the younger members of the group. Our ancestors had begun to handle objects and transform them with a view to their future use. They had started to build a mental model of their creation. The earliest humans identified the sites near rivers where they could find the best stones to use at butchery sites. So they had strong aptitudes for social organization, foresight, and planning. All of these skills are found in the parietal and frontal lobes of the brain, which in fact got bigger subsequently.

This process was once again not linear, though. *Homo habilis* showed great internal variability from individual to individual (with brains ranging from 600 to 800 cubic centimeters, or 36 to 48 cubic inches), and was perhaps not the only early species in our genus. For a long time, the new skills in technological design remained highly stable and were not accompanied by any great changes in other behaviors. Those tools were probably more than enough to satisfy the needs of the time. But if the technologies were so advantageous and the brain was galloping toward perfection, why was there so much cultural conservatism?

Perhaps the selective drive toward big brains is to be found elsewhere. Not in active hunting, however, because humans long remained opportunistic scavengers of the savanna, alongside hyenas and vultures. And not in their habitat, since that was the same as other hominins that survived up to 1.5 million years ago. Nor in geographic mobility, if it is true that we can already find human beings outside Africa, in the Middle East and Georgia, 2 million years ago who were able to survive in unfamiliar environments despite their modest brains. The remaining idea is that our encephalon increased in parallel with the growing complexity of social relations and a slowing down of individual development (another crucial and imperfect phenomenon that we will discuss later) that favored its plasticity as well as capacity for learning, imitation, and creative innovation.

The fact is that the brain expanded in one way or another in various species of the genus *Homo* (more than twelve have been identified to date), and this favored the development of abilities in observation, visuospatial skills, mental association, social coordination, and interpretation of the signs left by prey and predators. Yet this advance was neither smooth nor a foregone conclusion. An ancient relic of the first expansion of humans out of Africa survived for at least a million years on the island of Flores in Indonesia, becoming extinct only fifty thousand years ago. As a result of a selective process known as island dwarfism, which reduces the size of larger mammals stranded on islands, *Homo floresiensis* shrank to become a pygmy human species, with an average height of one meter (one yard) and a brain size of 420 cubic centimeters (25 cubic inches), or a third of ours. Nevertheless, they had developed a considerable amount of technology and were excellent hunters. They ate giant rats and dwarf elephants, and lived with Komodo dragons. On their own little island, they managed pretty well and might still be there had an invasive species called *Homo sapiens* not paid a visit.

So here we have a first exception to the principle regarding the inevitable growth of the brain in the genus *Homo*. At least one species went back to having a small head, but without losing much intelligence. Still, were such a case to be an isolated example, it could be treated as an extraordinary story that ends there. Instead, in 2015, another small species of the genus *Homo*, with a mix of archaic and modern features, was discovered, this time in South Africa near Johannesburg: *Homo naledi*. Thousands of bones were found, enough to suggest that its brain was 560 cubic centimeters (34 cubic inches) in males. But it was the dating, which occurred in 2017, that took everyone by surprise: not 2 million years as the morphology seemed to indicate, but between 335,000 and 236,000 years ago.

This scenario is surprising. In Africa, while some pretty large-brained populations of *Homo heidelbergensis* were transforming into *Homo sapiens*, there was a *Homo* with a brain one-third the size of ours wandering around South Africa, still well adapted to arboreal life, and not even geographically isolated. Soon thereafter, we also find the first Neanderthals in Europe, the Denisovans in Central Asia, and *Homo floresiensis* in Indonesia, with a brain variability between these species that ranged widely from 420 to 1,500 cubic centimeters (25 to 91 cubic inches). Now we have two examples of eccentric species of *Homo* with small brains that survived until recent times, disproving the old and well-established idea that there was a gradual trend toward encephalization in the genus *Homo* accompanied by progressive growth in social and technological complexity.

SO MANY COMPROMISES FOR A BRAIN

At an irregular pace, with combinations of traits and trajectories that differed from species to species, brain growth continued in some (though not all) lines of the genus *Homo*, culminating in the two brainiest cousins of all: we *Homo sapiens*, born in Africa between three hundred thousand and two hundred thousand years ago; and Neanderthals, who were indigenous to Europe for two hundred millennia (with deeper roots yet) and survived there up to forty thousand years ago. One of the many species, *Homo sapiens*, however, proved to be the carrier of an unprecedented mix of anatomical and cognitive traits that, much later on (not before seventy-five millennia ago), would transform it into a particularly flexible, mobile, creative, invasive, and talkative creature—so much so that today it remains the only representative of our genus, a late exception.

It seems that the expansion of the brain followed different routes in us and the Neanderthals. In their skulls, the brain mass gained space by expanding in a more antero-posterior direction and extending the head horizontally like a rugby ball. We, on the other hand, have expanded our brains in a more globular, football-like shape, by rounding the head in an upward direction and raising the forehead. The asymmetry between the right and left hemispheres, though, seems to have been common to the whole genus *Homo* from the outset, so this characteristic can reveal little concerning our specificity—at least as far as the rough dimensions are concerned. We know little about the internal organization of the brains of the past, except the fact that for us, the frontal areas (responsible for language, decision-making, and the processing of abstract concepts) and parietal areas became especially important. Yet even here, imperfect compromises are evident.

When one part of the totally integrated body grows more than the other parts, adjustments must be made. Our brain, which is characterized by considerable individual variation, is on average three times larger than that of an anthropoid ape with our same body size. So, over the course of two million years, the brain must have pressed against the skull walls quite a lot. Sometimes internal pressure prevailed over the shape of the skull, which adapted by sculpting itself to accommodate its increasingly bulky occupant. On other occasions, however, the structural constraints of the head (in relation to the face and jaw, for example) prevailed and the brain had to make do—an honorable compromise.

Another problem is that the brain is sensitive to temperature. Like the testicles, it should never overheat. But if it grows in volume, it automatically gets hotter because the surface area of dispersion does not grow cubically as its volume does but instead as the square. There is, in fact, a physical conflict between three- and two-dimensionality. The compromise was to generate a denser vascular network, whose functions not only involved oxygenation but also thermoregulation, similar to a car's radiator—another uncertain compromise, which is perhaps at the root of many headaches.

As is well known, this wonderful invention we carry in our heads is expensive; it burns 20 percent of our total energy. Nevertheless, considering our body mass, our metabolism as a whole has not increased compared to that of other mammals.

This means that in the course of our evolution we have had to offset the cost of our growing brain by saving energy elsewhere, perhaps in the digestive system, or perhaps by slowing down development, life processes, and related consumption. Undoubtedly, our diets became richer through taking in more animal protein (in the beginning, rotting meat and marrow stolen from carcasses—not very honorable, but that is how things went) and, in particular, by learning how to cook starchy tubers. Yet it is difficult to know whether this change in diet was a cause of brain growth or a consequence.

These digestive and dietary compromises must have paid off because the brain continued to push from within. The cortical cortex folded and folded back on itself to form a labyrinth of convolutions, but how this actually took place remains, unfortunately, purely guesswork. When the process came to an end, the greatest expansion had taken place in our parietal lobes, which are distinct from those of the other great apes and are significantly wider, longer, and more curved than those of the Neanderthals. This is something only we have. The upper part of our brain develops early at birth and is involved in various brain activities that it connects together. But it is particularly implicated in fundamental tasks in areas in which we excel, such as body management, visual and spatial integration, the agile use of the fingers, hand-eye coordination, and touch.

An ideal brain, however, does not necessarily have to have these characteristics. Starting from scratch, it could have been designed differently without both the clumsy array of parts and all those suboptimal compromises. Evolution is unintentional, and this is particularly evident in the brain with all of its quirks. Evolution adds irreversible variations to what already exists. According to some neurophysiologists, even our synaptic gaps are not as efficient as they could be. The other human species that survived until a few millennia ago—such as the Neanderthals with their brains that were as big as ours, if not bigger—were not "almost like us." Their brains were different (bigger at the back, in the occipital area, and less in the parietal lobes) and imperfect in their own ways. Simply put, our imperfection functioned better than theirs.

BRAIN TINKERING

Now let's try some generalizations. In the evolution by natural selection of this ingenious apparatus that we use to process and interpret the regularities of the world, two distinct mechanisms are at work, both of which generate imperfections (Marcus 2008). The first follows the so-called palimpsest principle—that is, it is analogous to a medieval parchment that was written on several times without ever completely erasing the previous text. We can clearly see this in action, for instance, in the layering of our old and new technologies. As we have already seen in numerous examples, evolution by natural selection makes a virtue out of necessity. It rarely throws away the old, substituting it with something completely new; it would be too expensive and dangerous because the need for survival is continuous. Similarly, in the brain we can see a permeating addition of new onto old, requiring coherent coordination between the old parts (usually associated with emotional and bodily functions) and new parts (associated with cognitive and linguistic activities). But the hierarchy between the two levels, and their separation, is by no means clear-cut.

So it happens that the ancient hindbrain that we share with so many of our animal cousins, and that controls basic functions such as breathing and balance, was juxtaposed with the newer midbrain, which is responsible for visual and auditory reflexes.

The latter was in turn supported by the evolutionarily younger forebrain, whose linguistic and deliberative functions are nonetheless not carried out independently from the older systems. Thus, the new has evolved by coordinating and integrating with the old so that the processing of evolutionarily recent tasks (such as speaking, imagining, and thinking) also requires the cooperation of older parts. It is not a graceful process, but it seems to have worked quite well in painter Leonardo da Vinci's head.

The second mechanism regards the evolution of new functions from old structures via the co-option of brain areas through the course of natural history. The human brain, as we have seen, began to grow about two million years ago, but only in the genus *Homo*. The survival functions at that time were different from today's. In the densely branched tree of our genus, the species *Homo sapiens* is young, having little more than three or two hundred millennia of history; it is a recent development in evolution, appearing when other human forms were already living on the planet. At first there seemed to be nothing special about it, but then a combination of traits proved successful. Starting from the previously accumulated evolutionary inertia, repeated readjustments were necessary in the brain of *Homo sapiens*, thereby creating space for further new potential reuses that certainly had not been foreseen in earlier phases.

Around seventy-five millennia ago, as usual beginning in Africa, modern human populations began to develop an array of behaviors and skills that had never been seen all together before, and were not directly linked to mere survival needs. Evidence for this includes objects engraved with abstract figures, body adornments, ritual burials, cave paintings, musical instruments, and technological innovations and variations. This was the dawn of imagination—a late dawn, but a rapid one. Considering the timing, it is likely that, as was the case with DNA, certain areas of the brain (especially in the frontal and parietal lobes) also expanded and duplicated in such a way that one part could continue to perform the original basic function while the other could specialize in more advanced skills.

The result is that our brain, thanks also to these imperfections, is plastic twice over because its central adaptive quality is ductility in adapting to environmental stress. First, the neural circuits are able to acquire functions for which they were not programmed in the course of evolution, and they can do this quickly and flexibly. We have been reading and writing for only five thousand years, and our brains certainly did not evolve "for" such tasks, yet they perform them quite well. The plastic coordination of neural development makes it possible to adapt and expand certain areas at the expense of others during growth, offering great advantages. So the brains of those who read and those who never learned to are brains that are not only culturally but biologically different. Using one technology rather than another sculpts the functioning of our cortex differently. This means that biology and culture influence each other.

Second, in the course of evolution, circuits that were initially dedicated to certain functions (e.g., of a sensory-motor nature) were co-opted for different ones (e.g., the use of tools, communication, and abstraction) as the environmental and social context changed. The areas related to manual dexterity and gestures are closely intertwined, for example, with those devoted to language, which suggests an evolutionary link or even an overlap between the two functions. Neural migrations, compensations, restructuring, unexpected conversions, and neural recycling are the processes that make our brains so plastic in both their development and evolution.

Today we know that the evolutionary remodeling of the human brain was genetically based, as we saw in the previous chapter with the osteocrin gene that was co-opted from the muscles to the brain. This is evident in all of its uniquenesses as each of us grows. An important distinguishing characteristic of the human brain regards its slow development, which takes almost two decades following birth. Childhood and adolescence, as we will see in the next chapter, represent our most important evolutionary heritage. This means that human cognition is particularly dependent on experiential learning. Through selective reduction, migration, differentiation, and other processes, neural circuits undergo profound structural and functional changes during growth. They are literally sculpted by our experiences. During our development, these changes in the brain are mediated by genes whose transcription is regulated by neuronal activity itself. It is genes, such as that osteocrin gene, that make our brains a masterpiece of plasticity.

This evolutionary medal, as is often the case, comes at a cost. Our brain is a frozen accident of evolution that was not designed from scratch. For this reason, it falls ill and loses control easily. Given this negative evolutionary characteristic, rather than looking for improbable hidden benefits in some human mental disorders as some scholars are prone to do (as if everything has happened for the best), it is much simpler to admit that imperfection takes its toll on us in the sad and countless forms of mental suffering. Schizophrenia and depression are not ancestral adaptations that have gone wrong. They are problems. In much the same way, it is possible that imperfections in our plastic and expanded brain increase our vulnerability toward certain degenerative diseases. As Levi-Montalcini (1987) pointed out, our brain is the outcome of disharmonious processes, related to psychic complexes and behavioral anomalies.

We can now add what we have learned in the previous chapter. Two of the most complex and creative systems ever invented by evolution—the genome and the brain—are reticular, redundant, clearly imperfect, and unnecessarily complicated. They are the results of adjustments, improvisations, and compensations. Neither would pass an engineering test. They look like the machines invented by the Californian cartoonist Rube Goldberg: incredibly complicated and far-fetched machines, the result of a chain reaction of causes and effects, which carry out operations that, on the whole, are simple or even superfluous. Nevertheless, these two inelegant systems are the most extraordinary things that nature has produced on this planet in 3.5 billion years. They are the two biological sources of pride we would happily put on display for a well-meaning alien visitor.

It is almost as if evolution has installed a Ferrari engine in a wonderful old racing car from the last century. The effect is extraordinary, but it is logical to presume that not everything works perfectly afterward. Again, in Jacob's (1977) words, "The formation of a dominant neocortex, preservation of an ancient nervous and hormonal system, which has remained partially autonomous and partially placed under the protection of the neocortex: this evolutionary process is very much like tinkering." Equally predictably, the intellectual and behavioral performance of such a contraption can be ambivalent: in some respects, amazing, and in others, horrific. *Homo sapiens* itself, displaying that symbolic and imaginative form of creativity, soon became an intrusive and overriding presence (Tattersall 2012), driving all other human species to extinction and profoundly altering the ecosystems they came into contact with.

US AND THEM

The two mechanisms described here (palimpsest and functional co-optation) created an evolutionary discrepancy in our brains, an intrinsic disharmony. The new parts in the neocortex were not restructurings of the older ones because the latter continued to perform basic biological functions required for the survival of the individual and therefore were well protected and preserved by natural selection. The new areas rode on the back of the old ones, and in some way managed to cope with disparity in the functions performed by neocortical areas and the limbic system, accentuated by the accumulation of cultural heritage (Levi-Montalcini 1987). In a moment, we will return to this potentially devastating evolutionary divide that lurks within our brains.

So brain plasticity is also a two-faced Janus. On the one hand, the malleability of our minds makes their indoctrination simple, and cultural factors can literally shape the conduct of individuals and entire crowds, even leading to the most pernicious of outcomes. On the other hand, early education regarding the values of civilization can deactivate and repress our more negative instincts, which are not as binding and invincible for us as they are in other animals. An example taken from neuroscience can illustrate this duality well.

According to several recent studies of visualization using functional magnetic resonance imaging, when our brain is exposed to the faces of foreign people—belonging to other human populations with different physiognomies (what were once erroneously called "human races")—it displays a fascinating and contradictory reaction. If a white person is shown the face of a Black person, or vice versa, deep subcortical areas, in particular the amygdala, are immediately activated, signaling a potential threat. The brain seems to be saying, "Who is this? They are not part of my community, they are unusual, they are not one of us." But this unconscious perception is short-lived because almost immediately the higher cortical areas enter into play, contradicting and regulating the original automatic emotional reaction, and so another cortical area reconciles the first two. It is as if a voice of reason and self-control has entered the scene to restore calm, making us aware that this is just another human face.

In practice, scientists have recorded an internal conflict in the government of our mind. It's a conflict between immediate negative impulses and egalitarian intentions, between implicit and explicit attitudes—a conflict that seems to have a precise evolutionary reason. As Darwin had already hypothesized and a lot of data have since confirmed, the species *Homo sapiens* comes from a long history of small social groups. Our strength consisted in being part of a small, well-organized, cohesive community, which was internally united and almost always in conflict with other tribes. So, paradoxically, conflict (between groups) led to the birth of altruism (within our group) (Bowles 2008). This led to our strong propensity toward immediately classifying someone as being, or not being, one of "us." It was important to make this distinction, and to make it quickly. In this attitude, it is easy to see the ambivalent roots of cooperation, on the one hand, and of conformism, tribalism, and sectarianism, on the other.

The present-day heritage of this history is that neural regions that have different evolutionary histories come into conflict when we put them opposite each other. Yet, when all goes well and there are no other constraints, a compromise can be found from time to time. The first lesson to be learned here is that if the cultural and educational contexts, propaganda, and social stereotypes that surround us while we grow up encourage discrimination and a fear of diversity, we will develop a natural tendency to take refuge in a protective "us" and see the "other" as a danger. It is a latent tendency, hidden just below the surface, and with a little indoctrination and propaganda, it surfaces again and can provoke considerable damage. In recent history, for example, think of how successful the criminal and intentional depictions of diversity as the enemy have been: more often than not, they resulted in massacres and ethnic cleansing.

If someone can maliciously take advantage of this imperfection by infiltrating our implicit preferences, it is fortunate that the opposite can also take place. In the experiments mentioned above, it is evident that cultural and social learning can greatly mitigate instinctive reactions (Kubota, Banaji, and Phelps 2012). If, for instance, the other person's face is that of a famous athlete or singer, the amygdala is not activated because we immediately recognize them as familiar, as "one of us"—which proves that individual experience, culture, and education count for a lot. The differences in reaction to otherness observed within each group, especially between Black and white people, depend very much on the individual histories of the subjects. This means there are antidotes to prejudices and "gut" reactions.

The problem is not only that there are today people who speculate (successfully) on the worst human prejudices, playing with fire and the danger of history repeating itself, but that our minds are objectively challenged by the enlargement of our historical "us," which is currently becoming increasingly wide, more metropolitan, more global, and frayed. It is an "us" that, for less than a hundred years, important international documents on universal rights have identified as the human species itself. In the tug-of-war within our unbalanced minds, that "us" can be fascinating but also frightening, and so we pull back and take refuge in our old and crazy tribe, whether it be real or digital (Cavalli-Sforza and Cavalli-Sforza 1995; Cavalli-Sforza 2000). As can be seen in many news stories, while we are planning to go to Mars, the limbic system that directs our emotional universe is still the limbic system of a primate. Advertisers and demagogues know this well.

6

THE IMPERFECT SAGE

> In the neighbourhood, there happened to live a very famous dervish, who was reputed to be the finest philosopher in Turkey. They went to consult him. Pangloss acted as their spokesman, and said to him: "Master, we've come to ask you to tell us why such a strange animal as man was created." "What's that to do with you?" said the dervish. "Is it any of your business?" "But, reverend father," Candide said, "there's an awful lot of evil in the world." "What does it matter if there's evil or good?" the dervish replied. "When His Highness sends a ship to Egypt, does he worry if the mice on board are comfortable or not?"
> —Voltaire, *Candide, or Optimism*

If DNA and the brain have not convinced you, let's have a look at the alleged perfection of the human body. From the nucleus of each cell to the architecture of our organs, the human body also represents a time capsule, which bears the traces and wounds of a long and contrasting evolutionary history. Naturally, not everything in our bodies serves a purpose, otherwise we would have to ask ourselves why we (and not the Neanderthals) have chins. To look for a function at all costs would be ridiculous. Pangloss could have said that a chin is needed in order to grow a goatee. In reality, the chin is not a distinct trait but rather the accidental product of the interaction between two developmental processes (alveolar and mandibular) of our face, which has flattened and softened over the course of evolution.

Despite the beautiful proportions of da Vinci's *Vitruvian Man* inscribed within a circle and square, our physique is mainly a compendium of mismatches worthy of Homer Simpson. We think the sunfish and the long-eared jerboa look strange and clumsy, but they must have the same opinion of us. I have already listed some vestigial features of the body of *Homo sapiens* in chapter 3. They are evidence of evolution (Williams 2006; Lieberman 2013). In males, what is the point of the urethra passing right through the center of the prostate, whose function is in no way linked to urination? The result is that, when the latter becomes inflamed and enlarged over the years, there is a lot of unnecessary pain. This makes absolutely no sense other than the fact that until recently, people did not grow old enough to suffer from such an ailment. It makes no sense, but this is evolution. Many of the other ills and pains of old age are only understandable if we assume that selective processes have taken no interest in humans after they have passed reproductive age. The tiresome imperfections of old age are transmitted because their carriers develop them after they have already had offspring.

THE MOST IMPERFECT OF REVOLUTIONS: WALKING

Let us now return to our second law. We have said that imperfection in nature arises from the need to find compromises between different needs and antagonistic selective drives. This means that an advantageous trait can evolve and succeed despite the fact that its owners pay the price in the form of annoying side effects. The cecal appendix is a human vestigial trait that was discarded due to the reduction of the intestine as a result of a change in diet (more diversified and no longer dominated by vegetables alone) in the genus *Homo*. According to recent studies, it may have some secondary advantages related to the immune system or could act as a reservoir of good bacteria in case of infections. This does not mean, however, that there are no permanent disadvantages such as a high rate of obstruction and consequent infection, which claimed many lives before the invention of surgery. There were numerous potential anatomical solutions that would have been more efficient than this one. Today, when needed, we can get by through an emergency operation, but to have a wormlike appendix in our bellies is decidedly a bad idea.

As we have seen, concealed ovulation is another oddity that we have almost uniquely. Human males do not perceive the moment when females are ready to produce offspring. In the more reasonable baboons, mandrills, chimpanzees, and bonobos, the female in estrus is recognizable due to the appearance, turgidity, and coloration of her genitals and the emanation of specific odors. This means that even the most obtuse male will sooner or later understand when it is time to do his duty, while in the human species, this does not happen. For us, and a few other species such as the gray langur of Southeast Asia, ovulation is concealed. This all goes to produce a great sense of insecurity in the males, who do not know if their copulation, which is often obtained at a high price, has been successful or not.

Considering how expensive sex is, it would be much better to coordinate this activity and avoid all of this effort. In fact, this is the norm in nature. But evidently the females of our species had to cope with males who were a little overpromiscuous. They needed a strategy to hold onto them. By making them perpetually insecure about their partner's fertility, they forced them to stay close to them all the time, watch over them by mating several times during the menstrual cycle, and possibly contribute to child support (this scenario, of course, applies to our progenitors, not to contemporary human societies, for the reasons we will see later). For males, this was not the ideal solution, but let us console ourselves because it is by no means the worst of the aforementioned imperfections of sex.

There is a long list of oddities. If you can move your auricula like elephants, it means you have useless but still-functioning muscles in your ears. The caudal vertebrae that are fused together under the pelvis are in fact the remnants of your tail, and the coccyx still serves as a point of attachment for some muscles. Just ask the opinion of anyone who has fallen down the stairs and violently hit the edge of a step. Those who suffer the debilitating back pain well understand the imperfections of human bipedalism, a compendium of locomotor inefficiency that made us human. For a moment, let us consider our strange posture.

The human spine did not evolve out of thin air. The supple spine of a quadruped or brachiator (the preexisting constraint and evolutionary inertia) was straightened out, meaning that the weight of the whole body now rests on a single axis and offloads on the two legs. As a result, the spine is curved and the vertebrae are subjected to undue pressure. Nerves and muscles have readjusted themselves as far as possible, but not enough to prevent sciatica, hernias, and flat feet. If, after all that effort made to stand upright on its lower limbs, that biped spends all of its days sitting at a desk or in a car, however, we are actively in search of the pain of imperfection.

So why become bipedal? This question is actually more difficult than it sounds. The upright posture is said to have enabled long-distance running and greater flexibility in locomotion. A quadruped will destroy us over a hundred yards, but we perform well as cross-country runners. As bipeds we can climb if necessary, as well as walk, run, and ford a river. This may well be true, but the fact remains that in the open spaces of the savanna, the other animals are almost all quadrupeds and so far they have coped quite well. It has also been said that bipedalism offered our ancestors the possibility of showing themselves to predators as a vertical rather than horizontal silhouette (much more visible to the average feline) and of standing above the grass to better spot predators at a distance. And then, of course, an upright posture freed our hands and arms from locomotion so they could be used to manipulate tools as well as carry food and the young. But did we become bipeds to free our hands, or did our hands become free because we became bipeds?

Alas, the numbers do not add up because the first lithic technologies appeared in Africa as far back as 3.3 million years ago, when there were still 700,000 years to go before the arrival of the genus *Homo*, and as far as we know only the australopithecines and *Kenyanthropus* with their many arboreal traits were still roaming around Lake Turkana. Why did technologies come first, and then total bipedalism? What is the cause and what is the consequence? Remembering of course that the bipedal gait costs us all of those expensive imperfections listed above, it must have been worth it right from the start, as otherwise our more arboreal counterparts would have prevailed.

Other experts in human evolution maintain that the initial push toward bipedalism was linked to thermoregulation. When species that live in zones bordering between forest and grassland explore open sunny areas with no shade, they have a serious problem regarding the maintenance of body temperature within certain physiological limits, and this applies particularly to the brain, which, as we have seen, does not tolerate overheating well. Savanna quadrupeds have developed appropriate countermeasures that are lacking in hominins like us. The solution adopted by our tribe appears to have been the reduction of surface area exposed to the sun to ensure that our body temperature was kept under control. At the same time, our ancestors may have gradually lost their fur and developed sweat glands. If this is the case, a thermoregulatory adaptation could then have triggered the cascade of advantageous uses (flexible locomotion, liberation of the upper limbs, etc.) that made bipedalism a good strategy despite its costs.

It is also likely that, due to these compromises, and despite its undoubted merits, bipedalism evolved slowly and timidly over a period of four million years, following several failed attempts and unsuccessful experiments. One such example is *Ardipithecus*, which was a forest biped that walked along branches. For two-thirds of the natural history of hominins (six to two million years ago), our ancestors, cousins, and relatives rightly preferred a hybrid solution: an arboreal life so they could protect themselves from predators (with persistent ancient traits such as curved fingers and long arms) and the prudent bipedal exploration of open glades in search of food. Lucy lived in this way, and died when she fell out of a tree. This was by far the most intelligent strategy at the time for those that were yet to become brave hunters, but were delicious prey for felines and giant eagles. Today, baboons and many other primates do the same. So let us forget the story of human evolution that begins with the heroic "descent from the trees" to conquer the savanna on foot. Only in the early days of the genus *Homo* did we become complete bipeds.

And many of our companions still curse that day. Walking upright on your legs becomes a big risk if your diet changes in the meantime, your brain starts to grow, and you have to give birth. The pelvis cannot expand much because if it did, you would not be able to stand upright. Consequently, the baby's head passes with considerable difficulty. If you could reset and go back, the ideal engineering solution would be to give birth directly from the abdomen, but this is not possible because our birth canal is a modified version of that of reptiles, which lay eggs, and of early mammals, which give birth to tiny offspring via the pelvis. So compromises are improvised, fixing pregnancy at nine months and giving birth to helpless babies whose brains are only one-third developed, with the remaining two-thirds being completed later. It remains a truly imperfect solution, however, if we think not only of how many mothers and babies have died during childbirth, but of how painful it is for women at the best of times.

The transition to bipedalism generated negative consequences in almost every part of the body. Human feet, with their plantigrade locomotion, have to tolerate high stress levels. Our neck, with that heavy, swinging bowling ball balanced on top, becomes a weak point. The abdomen, with all of its internal organs, is exposed to all sorts of trauma. The peritoneum is being pushed down by the force of gravity, provoking a predisposition to hernias and prolapses. You might even feel the consequences on your face. The next time you have a cold and feel the mucus pressing into every orifice of your face, think about the fact that your constipated maxillary sinuses have their drainage channels pointing upward toward the nasal cavities—against gravity! This makes them completely inefficient and easily clogged up with mucus as well as with other slimy substances. This seems like a bad design, but the fact is that in a quadruped, the opening of the maxillary sinuses faces forward, which works well. Yet for former quadrupeds like us, our faces have only recently adopted a vertical position, and this is the result.

Archaeologist André Leroi-Gourhan (1964) was right in saying that the history of humanity began with good feet, before great brains. But it was an ordeal, particularly in the beginning. Then we grew to like it, and with those legs we became migrant primates, with a strong sense of curiosity and no more boundaries to hold us back.

HOW TO TURN FRAGILITY INTO STRENGTH

We also became primates whose strange ambition was to grow old as late as possible. A wonderful ambition, but with it came a rich variety of imperfections too. Primates are the mammals that have the slowest and most delayed development. Instead of jumping up onto their feet in order to run after the mother, as all herbivores do to defend themselves from predators, monkeys like us remain babies longer, being protected by the group, and this guarantees a prolonged period of social learning, playing, and training in preparation for the hardship of future life. Like bipedalism and exorbitant brain growth, this is a costly and risky adaptation that requires adjustments as well as compromises that balance costs against benefit. Changing developmental rhythms is a common life strategy (even ostriches have wings and fluff when they are chicks), but it must be well balanced.

There are many effects generated from this delayed development: babies are born completely defenseless and totally dependent on care; their brains mature later and mainly after birth; childhood and adolescence are disproportionately prolonged; sexual maturity is reached later; and longevity, in general, increases. In short, life is enjoyed at a slower pace and no longer in a frantic race similar to rodents that live only a few years. Another effect of this change (called neoteny) is that adults tend to retain some juvenile traits. The features of a baby chimpanzee (a flatter face and rounder head) seem a little more "human" than an adult because we have retained these infantile characters throughout our life. We look like chimpanzees that have never grown up.

Of all primates, the genus *Homo* is the most neotenic. This speciality has become our trademark. We are the monkeys that remain children for the longest time of all. Human infants are practically helpless when they are born, but later become learning prodigies. Paradoxically, this delay in development has become our trump card. We have transformed a vulnerability, another imperfection, a fragility, into a strength. But frailties remain given that our children require much longer and more meticulous attention, specifically, a much greater investment in parental and social care.

These are all energies that adults subtracted from other activities at a time when we were still threatened by predators that considered our babies an easy feast. Later, the enemies to be afraid of became rival groups of humans. The mothers of these young no longer had fur, so they had to carry their babies in their arms, keep an eye on them constantly, and keep them close through the use of recognizable sounds. In order to tolerate such a demanding change, human social groups must have already been sufficiently protective to ensure the survival of their young. This means that in the genus *Homo*, the selective pressures toward mere survival had weakened, thanks to sociality and formidable technological inventions such as fire. In fact, if life were too hard, you would not be able to afford to produce such slow-growing offspring because your main imperative would be to generate many children and ensure that those children outlived you.

If we then take a look inside the genus *Homo*—you will already have understood—we can see that *Homo sapiens* is the most neotenic species of all. A comparative paleontological study of dentine growth demonstrated that our species reaches sexual maturity a little later than Neanderthals. We are therefore the world champions of prolonged infancy (if we exclude the

Mexican axolotl salamander, which retains the characteristics of the larval stage throughout its life, and has an extraordinary ability to regenerate limbs and organs). The extension of our biological cycle, caused by mutations in the handful of genes that regulate development, has made it possible for us to invest more time in learning, imitation, play, and the curious exploration of the world, along with the transmission of ideas and technological skills.

It is also possible that one of the effects of neoteny involves favoring a larger brain in adulthood, given that the ratio of brain-to-body volume in children is higher than in adults. Cartoonists are well aware that if they want to generate feelings of tenderness in their viewers, they have to draw their characters with a big, round head, two big eyes, and a small body. The great ethologist Konrad Lorenz had already noted that children's morphological characteristics trigger innate mechanisms of care and protection in adults, thereby suggesting that maintaining the compromise between parental investment and cultural learning requires a deep-rooted instinct. The softening of *Homo sapiens* traits seems to have been further facilitated by a process of "self-domestication"—that is, natural selection favoring more docile and sociable individuals.

As a result, and as we have seen, our brain has become extremely plastic, to the point that it is practically a sponge of information that lets itself be sculpted by experience. But here, too, we can find imperfection. For example, one unfortunate side effect is that childhood traumas and deprivations in the early years of life can leave an indelible mark on our minds. Yet the advantages of neoteny eventually prevailed, or otherwise today we would have children already enrolled in primary school at the age of two and with the ability to speak at the age of six months. In fact, one of the beneficial effects of neoteny regards the extension and refinement of the acquisition of one or more mother tongues—one of the many historical and cultural versions of the articulated language with which *Homo sapiens* is endowed. According to some paleoanthropologists, the very evolution of language, with its free play of arbitrary associations between sounds and meanings, can be linked precisely to the vocal experiments of children during that protected phase of life in which they did not, like adults, have to worry first and foremost about simply surviving and then conversing.

But don't fall into the trap of thinking that the wonderful language of humans is perfect.

SORRY, CAN YOU REPEAT THAT, PLEASE?

In order to speak, we have to assemble an almost comically complex piece of machinery through which we can inhale and exhale, and then we must emit a flow of sound, modulated by the vocal cords, through the vocal tract and its many components (glottis, palate, tongue, teeth, and lips). This contraption, which might well seem sublime to some, actually limits us to a range of about ninety sounds, and little more. All the languages in the world must work with this limited array of phonemes, and in most cases they use far fewer. But this is not the only imperfection. The cognitive control of language is carried out obviously by the brain, whose problems we have already mentioned. Furthermore, our language was made possible by the lowering of the larynx, a trait that also occurs in other animals, but with completely different functions.

In male deer, the lowered larynx is used to give a false impression of size when competing with other males in mating: a lowered larynx suggests that the animal is bigger than it actually is, and misleads other males and females. This has nothing to do with language at all. Much later, when *Homo sapiens* was born in Africa, we find a species with a slender physique and decidedly elongated neck (Neanderthal had a more flattened one, perhaps as a form of protection against the cold). With the particularly imperfect adoption of bipedalism, the larynx moved down into the neck and split the vocal tract into two almost perpendicular sections: vertical and horizontal.

In the meantime, this lowering had become useful in functions that were different from the cheating deer. With this superior laryngeal conformation, we can produce language, but, as Darwin had already pointed out, it exposes us to the serious risk of choking: any piece of food or drop of liquid that we swallow passes close to the opening of the trachea—not a good idea. It's not just a nightmare, it's an imperfect reality that continues to kill thousands of people every year, despite the emergency rescue maneuvers we know today.

So language is yet another expensive adaptation because we pay that potential price of choking. It must have been worth it, but it could have been achieved in more anatomically efficient ways too, eliminating the choking risk. Once again, it's a contrivance, a botched yet workable attempt to maintain two different functions: breathing and articulating speech. By recovering bits and pieces here and there during evolution, a digestive tube also became a respiratory tube, which in turn became a vocal apparatus. We could not reasonably have hoped for anything better.

So far, our considerations of language have been restricted to simple anatomy, but from a functional point of view, its actual effectiveness also leaves a lot to be desired. As geneticist Luigi Luca Cavalli-Sforza has often observed, there is nothing more imprecise, jumbled, redundant, and equivocal than human speech, not to mention its contingent and diverse historical manifestations, languages. Human language is powerful and allusive, capable of producing marvelous literary masterpieces, but it is also the source of countless and tiresome misunderstandings, arbitrary associations between words and meanings, vague generalizations, lexical ambiguities, and semantic incongruences as well as all manner of idiosyncratic irregularities. Besides wonderful theatrical plays, these imperfections have provoked divorces, broken friendships, and wars. The fact that language is so full of errors and gaps has led some to doubt that language actually evolved "to help" us communicate. Perhaps in the beginning, its function was simply to order our thoughts by picturing them in a form of uninterrupted inner dialogue.

The simple fact, however, is that now, even when humans speak the same language, they frequently have to understand each other according to the context of the conversation, which is just an elegant way of saying that we understand through trial and error. The arboreal recursiveness of syntax is obviously wonderful, and no other animal can do what we do, but when we encounter the third relative clause in a sentence, our memory will already be failing us, and we are lost. Still, were our expression to be crystal clear and infallible and without nuances, it would be incredibly boring. We would miss out on all the pleasures of puns, double entendres, witticisms, humor, irony, and cabaret. So let us enjoy the upside of such noncomputational suboptimality. And let us resign ourselves to the fact that emails and ridiculous emoticons can be misunderstood much more easily than a glance. But above all, let's stop getting angry if our car's voice commands don't always follow our instructions properly and ask us (politely) to repeat them. Put yourself in the shoes of a poor computer that must try and understand the irrational jumble of a human language!

ALWAYS A LITTLE LATE

Let's do some math. Apart from DNA and the brain, which have their own problems, we humans are the way we are thanks to three significant imperfections that have equally made our fortune: bipedalism, neoteny, and language. Thanks to these sublime compromises, humans have changed a lot in the last two million years. But what if the environment around us has changed even more rapidly?

In evolution, sometimes adaptations that contributed positively to the survival and reproduction of a population under ancestral environmental conditions can turn into maladaptations as a result of profound and rapid changes in the environment. You have no time to reorganize things, and you find yourself lagging behind, out of phase. In fact, there are considerable differences between the world we live in today and the habitats in which our small bands of human hunters and gatherers evolved for two hundred millennia. Indeed, it is often human technological and cultural activities that have radically transformed the environment around us. For example, consider fire, cooking (today we can no longer survive on raw food alone), the digestion of milk even in adulthood, and the spread of fermented alcoholic beverages.

This delay in evolution can cause serious problems. For instance, if our digestive system evolved over a long period of time in an environment in which food was scarce and uncertain, its adaptation will rightly consist of trying to absorb as many calories as possible once it finds a source of nutrition in order to store sugars and fats for as long as possible until the next uncertain supply arrives. Days could go by between meals. Better to grab what you can get while you can. Yet such an adaptation becomes counterproductive if individuals suddenly find themselves in a world of fast junk food, surrounded by aisles overflowing with cheap food that is full of fats and sugars and temptingly, if bulkily, packaged in harmful and useless plastic. In the space of just a few millennia, food supplies have changed from being "scarce and uncertain" to "abundant and continuous" for a significant proportion of humanity. Obesity can therefore also be traced back to a recent and excessively rapid calorie enrichment of the diets of part of modern humankind that is ill-suited to the microbiota and metabolic processes that evolved slowly in ancient times.

Today this environmental maladaptation hypothesis is being used to help us understand the genesis (and perhaps treatment) of other serious diseases such as diabetes, heart ailments, allergies, myopia, and autoimmune syndromes. In the past, many fears and anxieties had an adaptive value because they warned us of potential dangers. Unfortunately, this value has largely been lost today. So we are left with the cost of suffering, but without any quid pro quo. And it is above all in the dynamics of human pleasure that this process reveals all of its intrinsic capacity to generate imperfections.

According to biologists, pleasure has evolved as an incentive and guide toward behavior that favors survival and reproduction, especially if it requires particularly heavy investment as in the case of sex and the subsequent maintenance of offspring. If sex (a producer of diversity) is to persist and not be substituted by more straightforward and less strenuous reproductive methods (such as cloning), it is important that its practitioners be rewarded. In other words, pleasure. But the game can go beyond its initial constraints. Today we pursue culturally constructed yet biologically gratifying pleasures that have nothing to do with their original function, such as the pleasures of sex regardless of any desire to reproduce.

We can naturally welcome this extension of pleasure. Sometimes, though, the side effects of essential pleasures, such as sociability and curiosity, can become dangerous. We have certainly not evolved to sit for hours in front of the television, to gamble, or to stay glued to a computer screen night after night, immersed in the banal chatter of a social network or a competitive trance in front of the mental fiction of a video game. Plants have devised a whole range of psychotropic substances that take advantage of natural imperfections in order to break through the pleasure systems of animals. And we, too, become addicted to drugs and stimulants. We are vulnerable because the roots of pleasure persist but have lost their original function and thus go round and round in circles, looking for alternatives. All those years ago, our immediate need for a lot of sugar has today transformed into an uncontrollable desire for a sweet now, and at any cost, as an immediate reward. If one also considers that the environment around us is full of glittering and sensational messages that stimulate those hidden chords, the harmful consequences come as no surprise.

This further mismatch between historical origins and current function (or nonfunction), and between primary function and subsequent environmental changes, generates a condition of maladaptation that we could define as the sixth law of imperfection: *when the environment runs faster than we can, we find ourselves evolutionarily out of phase, and therefore a little unsuitable or imperfect*. Once again, when considering imperfection, we realize from another point of view that evolution is a constant struggle between the available material—such as constraints, variability, and past history—and the ever-changing environment around us.

In the second half of the 1970s, an evolutionist at the University of Chicago, Leigh Van Valen, coined the so-called Red Queen hypothesis, named after Lewis Carroll's passage in

Through the Looking Glass in which the Red Queen forces Alice to run endlessly and ever faster in order to remain in the same place. The idea is simple: natural selection does not necessarily lead to an accumulation of positive adaptive experience because environments change randomly, unpredictably, and sometimes so fast that organisms are forced into a potentially endless adaptive race to catch up. Just like the Red Queen, the selective game continuously transforms organisms to help them keep up with environmental changes. And sometimes they can fall behind.

The environment of one species can also be another species, and so evolution becomes a duet or dance, such as in the reckless arms races between prey and predators, between hosts and parasites (like viruses), between plants and pollinators. If we study and photograph this process of dynamic equilibrium at one specific point in time, one participant may be at a disadvantage compared to the other. Someone is succumbing because they have yet to come up with a countermove (e.g., an immune response, natural or acquired through a vaccine, against a new pathogen). The result, according to Van Valen, is that organisms are often a little maladjusted, a little behind in their environmental niche, and therefore in a situation of suboptimal adaptation. In short, regarding the specific conditions of their environment, they are always a little imperfect.

We humans are late twice over: in our development as well as in our adaptation to the changing environment that we ourselves have transformed. If we create anthropic niches that favor the spillover of viruses, then we have to adapt, with major efforts and a high price, to a world in which pandemics are becoming more and more frequent. For the same reasons, we feel inadequate when we see that our children are natural natives of technologies that are alien to us. Kids are using those two opposable thumbs in a way that evolution could never have imagined over the last six million years. But this process is ongoing and applies to the past as well; it is wrong to presume that all of our ailments are linked to the stress of modern life and that the good old days were a harmonious Garden of Eden.

There is no reason to believe that the present imperfection (of the sixth kind) did not exist in the past. Similar to the nostalgic Uncle Vanya in Roy Lewis's (1960) *The Evolution of Man*, in the Paleolithic period there must have been old conservatives who spoke out against the innovative impulses of those who tried to rebel against nature by inventing fire, arrows, spears, marriage, exploration, and other devilry. "It would have been far better to stay in the trees!" they probably said, while the world around them had already changed.

7

WOULD YOU BUY A SECONDHAND CAR FROM A *HOMO SAPIENS*?

> Nothing could have been so splendid, so elegant, so dazzling, so orderly, as the two armies. The trumpets, fifes, oboes, drums, and cannons produced a harmony as was never heard in hell. First the cannons felled about six thousand men on each side. Then the musketry removed from the best of worlds between nine and ten thousand rogues who were poisoning its surface. The bayonet, too, was good enough cause for the death of some thousands of men. The total could well have amounted to some thirty thousand souls. Candide, trembling like a philosopher, hid himself as best he could from the heroic massacre.
>
> —Voltaire, *Candide, or Optimism*

Geneticist Theodosius Dobzhansky believed in the improving and refining power of natural selection, and thought that human imperfections were due to the young age of our species. He considered us to be a recent product of evolution and therefore to have had little time to make the necessary biological refinements. This would suggest that our limitations and flaws are not due to the decadence of an exhausted and aging humankind. Lucretius also thought that the world was at its beginning, and thus still to be modeled and refined.

This is an intriguing thesis, but it must be said that in the short time we have been here, we have been particularly active.

We are young, brazen, and shameless. A cultural explosion has been grafted onto our biological evolution. We could therefore hypothesize that technological artifacts and inventions (these *are* designed *and* intentional, and hence potentially perfect) are the way in which we compensate for the innate biological limitations we have accumulated through the evolutionary tinkering we have been talking about so far. With the help of prosthetics, additions, and enhancements, today we can adapt the world to our needs rather than the other way around as it was in the past.

There is no doubt that our ideas, technologies, and social organization have helped us adapt to the entire world, enabling us to overcome what were previously insurmountable ecological and physical barriers. No primate has ever gone through such geographic expansion. As early as forty-five thousand years ago, groups of modern human hunters that had emerged from Africa a few millennia earlier were poaching and slaughtering mammoths in the Arctic, on the shores of the Kara Sea, and in extreme territories where even the mildest temperature was minus 25 degrees Celsius. Five millennia later, other human groups that were physically similar, but in a completely different hot and humid environment, painted and carved exquisite rock art on the tropical island of Sulawesi. From the North Pole to Indonesia, hunting and imagining, there were no more barriers remaining to stop us. Right from the get-go, creativity and invasiveness represented the two sides of our ambivalence. The experiment remains ongoing, and rather than an essence, humanity is still in the making. We are not a "being": we are a "becoming."

Yet our ubiquitous imperfection, far from being merely a complement to our incompleteness, is projected by us onto the technologies themselves. Efficiency is not the only criterion to be considered in evolution, and this is true for technological evolution as well.

THE BEAUTIFUL IMPERFECTION OF A KEYBOARD

To find perfection, we thought we had no option other than turning to engineering, design, and the marvelous technical abilities of *Homo sapiens*. No such thing. Let's take an example for a moment. When we picture a typewriter in our minds, we immediately think of the Olivetti Lettera 22 and its splendid shape, which is both beautiful and functional. We can imagine it on writer Ernest Hemingway's coffee table or exhibited at the Museum of Modern Art in New York City as a masterpiece of design, which in fact it is, having been designed in 1950 by architect Marcello Nizzoli.

The advertising poster for the Olivetti M1 typewriter, which was presented at the 1911 Universal Exposition of Turin, showed poet Dante Alighieri happily tapping away on the keys. Italian engineer Camillo Olivetti came up with the prototype after two trips to the United States, the first in 1892, and the second in 1904. Olivetti's initial idea was to combine technical innovation with an aesthetic sense of design, yet there was a hidden flaw.

It is unclear where this product originated, but it was developed to write standardized and easily legible texts and documents quickly by pressing the letters onto paper by way of an ink ribbon. Moreover, multiple copies could be produced through the use of carbon paper. At first it was considered by some (as so frequently happens with almost all inventions) as a threat because it transformed the art of writing into something mechanical and uniform. Goodbye handwriting! A "writing machine" had been invented by another Italian, Giuseppe Ravizza, in Novara in 1846; the aim was purely philanthropic because he wanted the blind to be able to write as well. His "harpsichord scribe" was patented in 1855, and the name represented the idea of inventing a machine with harpsichord keys, but that could be used for writing and not playing—in short, a piano for writing. Other real or alleged typewriter inventors, mainly Italian and German, claimed to have designed them to help the blind too. Yet none of these were contacted by a company wanting to produce them. The typewriters won a few medals at exhibitions and trade fairs, but nothing more.

Producing a typewriter on an industrial scale meant having to go overseas, where business sense was more aggressive and pragmatic. The prototype of the most widely used alphanumeric keyboard in the world today, the so-called QWERTY type, was patented by inventor Christopher Sholes in the United States in the 1860s. It was launched by Remington and Sons, a company that, until then, had only produced rifles and other weapons—with great success. By 1880, there were just over five thousand Sholes typewriters in circulation.

We seldom consider one particular detail of the old typewriters: the arrangement of the keys. Why are the letters arranged in that way? Lowercase letters on three rows with a single key for the uppercase letters. On the top row, from the left, the first six letters are QWERTY. This is strange given that in English more than 70 percent of words are written with other letters, which make up the DIATHENSOR sequence. So the most rational and efficient arrangement would have been to put these letters on the most accessible line, namely the second one, and preferably in a central position. In 1893 a keyboard with these specifications was in fact produced and marketed, but it was not successful.

Why was QWERTY so special and useful? It had a simple function, which was the technological equivalent of a selective pressure: to separate the most frequent letters from each other. The *A* and *O*, for example, are far apart, at the sides, and must be typed with the weaker fingers—that is, the ring and little finger (if you want confirmation regarding how often we use them, have a look at how quickly those two keys wear out and get dirty). In the middle row there are only some commonly used letters. For us, this is quite irrational and imperfect, but why? Because we live in a different world from the one where typewriters evolved.

One of the most important functions at that time was to distance the most frequently used letters from each other because typewriters had hammers that often overlapped and jammed, provoking delays and even physical damage. By spacing out the most frequently used letters, typing could still be reasonably fast, without jamming the hammers. The middle row is more or less in alphabetical order: DFGHJKL, but with the most frequently used letters, *E* and *I*, on the top row. So what had previously seemed irrational becomes not only sensible but also ingenious. This is how the QWERTY keyboard came into being, pragmatically and imperfectly: to slightly offset the quick sequential moments when the hammers struck the ink strip and pressed down on the paper. Excessive speed and an uneven typing rhythm would have been dysfunctional at that time; the hammers would have jammed together continuously, resulting in the same letter being typed several times over. This meant that typists had to stop everything in order to pick up an eraser or use some correcting fluid.

So typewriters were designed with the hammers and roller in view so that the problem of overlapping hammers could be immediately seen and sorted out, and with the most common letters well spaced out. Then, there are the pure contingencies of history. The *R* is not such a common letter, but it was added onto the top row on the left so that salespeople could impress potential buyers by typing the term "TYPEWRITER" using only the keys on the first line (try it and you'll see that it still works in the same way more than a century later).

But the QWERTY keyboard holds another secret that is much more difficult to identify. As early as 1932, a faster and more efficient keyboard, the DSK, which stands for Dvorak Simplified Keyboard, was launched and vaunted as the typing speed champion of the time. Nevertheless, the DSK failed to dislodge the QWERTY—the old, obsolete, and less efficient QWERTY. Today, it can seem even more surprising given that hammers were discarded decades ago and we have triumphantly entered the digital age, but our computer keyboards continue with the QWERTY sequence of letters on the left of the top row. This makes no sense.

Thirty years ago, Stanford economic historian Paul A. David tried to solve this riddle: Why has the QWERTY survived despite its clearly superior competitors? As early as 1872, inventor Thomas Alva Edison, no less, had patented an electric typewriter with a printing wheel. In 1879, the first typewriter without hammers, but with a rotating cylinder carrying all the letters, was launched on the market. Yet nothing changed; almost everyone went on using the QWERTY keyboard. Between 1880 and 1890, the world of typing had a wide variety of models to choose from, some without hammers. At the turn of the century, however, the QWERTY sequence became the industrial standard internationally. Rotating heads were invented, and in 1901, the first electric typewriter was manufactured.

Then IBM's Selectric model arrived on the scene, followed by the Hollerith punch card machine, and finally the personal computer keyboard—a momentous change, and indeed a revolutionary technological development that makes everything else seem like prehistory. Despite all of this, though, the QWERTY keyboard not only stubbornly persists but continues to dominate.

The QWERTY keyboard is a "frozen accident," a solution that was initially created for contingent functional reasons that soon became suboptimal, but that later established a dominant position, becoming a fashion, model, and norm thanks to a chain reaction of fortunate events. One of the first and most popular eight-finger typing schools, which opened in Cincinnati in 1882, used QWERTY keyboards. In 1888, a famous speed-typing championship was won by an expert typist who managed to type without looking at the keyboard—a QWERTY, of course. The first typing manuals used the QWERTY as a model and taught touch-typing. In short, a set of favorable yet unnecessary conditions led to the success of this model of typewriter.

History was channeled in that direction, and moving in another direction became costly and risky, despite all the inefficiency and imperfection. To date no one has taken the globally coordinated decision to replace all QWERTY keyboards with ergonomic alternatives. The operation could be economically successful, but it has yet to happen. As a result, almost all of us use an imperfect keyboard, or national variants, such as the German QWERTZ with a *Z* instead of a *Y*, or the French AZERTY. Naturally, perfection is not the only imperative of technological evolution. The QWERTY is the child of history and its contingencies. We are used to thinking that the fruit of human ingenuity is freer and less conditioned than organisms, so anchored to their DNA and physical constraints. Steel, glass, and plastic, on the other hand, are more malleable. True, today we have cars instead of horse-drawn carriages, and light bulbs instead of gas lamps. Technological evolution is quick and can be overwhelming. In the history of culture, we are the ones who choose an innovation, and we can immediately teach and spread it to everyone just by putting it on the internet. Not so in nature. In technology, we can replace and remix everything. Nevertheless, something in common with biological evolution remains.

Technologies, like living organisms, are not the product of optimal design but rather of a procedure that involves arrangements, imperfections, and functional co-options. Today, typewriters with their QWERTY keyboards have become a collector's item, and some people have started using them to give their writing a more authentic feel. What you are reading here is written with a QWERTY computer—that is, with an imperfect combination of two technologies born at completely different times and in completely different ways. And even this keyboard contains one of history's secrets, namely its contingency and imperfection. According to economists and technologists Kevin Kelly (2010) and William Brian Arthur (2009), many technological items have become well-established products regardless of their efficiency, and are the result of the reutilization of already-existing components and structures.

So even in technological innovation, nothing starts from scratch. Technological innovation usually involves a recombination and hybridization of existing technologies through new designs and uses. We begin with the material we have at our disposal and reorganize it; we move from preexisting constraints and modules, constantly changing, but starting with the creative reuse of what already exists. If Kelly and Brian Arthur are right, then evolutionary tinkering also applies to technological evolution and innovation.

Indeed, in the history of technology, it is rare to find an invention that has retained the same uses and functions as those originally intended by its creator. The phonograph was designed by Edison to do just about anything apart from playing recorded music. The same applies to the radio, transistors, lasers, the internet, and GPS—in other words, technologies (all the result of unpredictable scientific discoveries) that today permeate the environment where we are born, grow up, and live. This naturally also includes the multiple uses and extensions of cell phone technology, which is invading our lives with all of its trills.

THE EXPLOITS OF THE SELF-PROCLAIMED *SAPIENS*

Technologies are the way in which we change the world, but then it is the world that, with no warning, changes us. And so we near the final short-circuit of imperfection between our growing powers and ever-present limitations. I have already mentioned the latter, while for the former, a brief list will suffice. We have been evolving in tandem with technologies for three million years. We have domesticated plants and animals. After the Industrial Revolution of the steam engine and the exploitation of energy, and after the information boom, our lives are being populated by artificial intelligences, molecular machines, organoids, virtual reality viewers, bioprinters, human-machine interactions, synthetic bacteria, genetically edited organisms, robotic butlers, flying drones and spies, and devices that talk to each other on the net and give us a lot of advice that was not always requested. Our brains connect daily through the World Wide Web, the greatest experiment ever realized of a real-time collective manifestation of the best and worst of humanity.

In various laboratories around the world, lethal viruses can now be created that have been genetically modified to simulate their likely appearance in nature in new forms—a commendable objective, but with the risk of possible malicious use. Synthetic biologists have given birth to microbes with a minimal genome that have been assembled in the laboratory. Now they will adapt them through the addition of other sequences. So far, we have read the book of DNA, ours and that of hundreds of other species. Now we can rewrite it, copy and paste it, add and remove sequences. We can generate great advances in medicine, gene therapies, the synthesis of new drugs and vaccines, and the improvement of plants and animals, all of which are intertwined with insidious leaps forward such as that of creating gene-edited babies.

In the meantime, microplastics are spreading through the oceans down to the depths. Coral reefs are bleaching, seas are becoming more acidic, and the polar ice caps are receding. Even the global warming that has been caused by human activities can be viewed as an enormous and highly risky experiment in environmental engineering on a global scale; the biosphere will get by somehow, though it's unclear whether *Homo sapiens* will do the same. We will come to understand one day that this senseless behavior interacts negatively with the terrible global inequalities that already exist, making them worse and thus generating famine, conflict, a sense of insecurity, and massive forced migrations. We change the world, and the world sends us the bill. It is called the "evolutionary trap," and it is not easy to escape from it.

The imminent danger will awaken consciences and minds, and who knows, the people born in the climate we have changed will find energy, transport, and building solutions that are unthinkable today. In the meantime, however, biodiversity is plummeting, and this includes animals that we presumed were resistant to just about anything, such as insects and other arthropods that are fundamental to the production of a large proportion of human crops. And we call ourselves "*sapiens*." The plastic desert we leave behind will be inherited by viruses, bacteria, jellyfish, cockroaches, rats, and scorpions, which, although not good company, are more perfect organisms than we are. Then, in our absence, the oceans and forests will return to their past glories. The causes of the current "sixth mass extinction" are well known: deforestation to make way for pastures, plantations, roads, and mines; the spread of invasive species through trade and tourism; relentless urbanization; agricultural and industrial pollution; and the overexploitation of resources through intensive hunting and fishing (Kolbert 2014; Wilson 2016). Our signature is scrawled across the planet.

Let us ignore our other minor flaws that can be seen on the surface of the globe, such as kamikaze terrorism, religious fundamentalism, ridiculous consumerism, conformism of all kinds, irrationality in our economic choices, political shortsightedness, populist incompetence, geopolitical lawlessness, predatory pricing, and corruption. Philosopher Michel de Montaigne said that every population is convinced that it has the perfect culture, perfect religion, and perfect government, and uses the supposed imperfection of others as a pretext to subjugate them. Nuclear power is in the hands of political classes that, more often than not, believe in the most varied forms of superstition and increasingly refuse to consider well-established scientific evidence. After all, we are excellent at hiding and eliminating inconvenient truths.

Even as we go through this long list of human villainies, pessimism does not prevail because the impression remains that our imperfect human nature has not reached the end of its journey but rather is just learning to walk, making its first uncertain steps, for better or worse. We are only at the beginning of it all. With microchips implanted in our brains, increasingly addicted to Google, hyperconnected to the cloud, and surrounded by biorobotic prostheses and synthetic organisms, whatever the prophets of posthumanism may say, we remain an updated and perfectible version of the good old African version of *Homo sapiens*. Naturally, our internal evolutionary gap will widen. The same large bipedal mammal that a hundred thousand years ago had to make do with sticks and chipped stones now drives spaceships, designs permanent lunar bases, invents nanotechnological marvels such as graphene, and predicts and then reveals the existence of the Higgs boson and gravitational waves. And so the circle of evolution and knowledge closes because one of our most dignified skills is our ability to look toward the boundaries of the universe and understand that we are the children of a cosmic sequence of imperfections and bifurcations, of Lucretian clinamen.

Assuming that we are imperfect beings who unconsciously build our own future, it is not worth challenging machines and artificial intelligence. Wherever there is a task that needs computing power, memory, precision, repetition, and speed, they will win. All that remains for us to do is to treasure their lack of imperfections because we program them, frequently to counteract our own imperfection. We have 3.75 billion years of rough-and-ready history behind us, including a number of fruitful accidents; they do not. Lacking the human magnitude of imperfection, they do not set out to write a poem or contemplate a work of art. They don't have that irrational twist that often got us out of trouble.

According to Levi-Montalcini (1987), imperfection is the dominant characteristic of the behavior of *Homo sapiens*, and it is clearly evident in both our historical tragedies and everyday miseries. As Montaigne (1958) wrote in his *Essays*, we are never entirely masters of ourselves and our impulses. Four centuries after, the human cocktail of growing powers and stubborn imperfections raises a fundamental question: If you were a wise god who was disinterested in this game, would you put all these weapons in the hands of a fellow such as *Homo sapiens*? Would you trust this anthropomorphic African ape who sprang up from nowhere two hundred millennia ago? Would you give them the keys to the spaceship we call Earth? Would you buy a secondhand car from a *Homo sapiens*? The answer depends on how much you think our species is capable of making rational choices.

FLAT EARTHERS ON A CRUISE

We have been saying for centuries that a human being is the number one rational animal, but then, as polymath Bertrand Russell noted, we spend our entire lives looking for some form of evidence to confirm this statement. And we find little. On the contrary, we find far more convincing evidence of our tendency toward self-deception. Whenever there is an accident, it is always everybody else's fault. We are masters at rationalizing after the event has taken place, making us good at excuses. Stereotypes drive us crazy. We love to presume that, when all is said and done, the victim of an abuse or injustice was in some way asking for it. We want everything now, even in situations where waiting would mean having more (this happens in the particularly imperfect period of adolescence, but not only). A greedy present is always worth more than a lucrative but uncertain future (Diamond 2005).

Some behavioral imperfections are specific to humans—that is, apparently we are the only animal in the world to have them. One is what the evolutionist Bill Hamilton referred to as the nonadaptive strategy of malevolence: harming others with no form of benefit for oneself, or in other words the ethological version of economic historian Carlo M. Cipolla's (1988) third fundamental law of human stupidity, which goes like this: A stupid person is someone who causes losses to others while he receives no advantage and may even experience losses. After all, what other mammal writes brutal trash on social media, lacking the courage to say it face-to-face, or backs up a leader who is clearly making a complete mess of things? In the world of humans, you can do harm without necessarily being a criminal. More often than not, damage is inflicted through sloppiness, ignorance, and sloth.

On the web, that sublime technological invention, we unleash our worst atavistic impulses and join comforting digital tribes among which we can give free rein to our football chants. Feeling nostalgia for the protective circle that we call "just us," we even look for primitive communities in the vast ocean of the net. Irrationally, we find comfort in those who already think like us and stubbornly look to confirm what we want to believe. We pretend that we are searching for information, but we have all the answers already. We uncritically accept pure nonsense and mystifications as long as they confirm our preconceptions, but we treat the ideas of others much more rigidly.

In the era of disappearing facts, which are being replaced by "perceptions," everything becomes interpretation, so angry screamers, along with professional and amateur conspiracy theorists, come to the fore. And we have collectively, democratically, and knowingly entrusted a great deal of power to these self-declared *Homo sapiens*. The misinformers of the web have evolved a common trait: they are all full of paranoid certainties, flatly refuse to take into consideration any counterevidence, and make fun of anyone who dares to challenge their nonsense. They are ordinary people, imperfect just like us, so we can vote for them without feeling inadequate.

We behave like know-it-alls, but we are constantly fooled day and night by an avalanche of advertisements designed to play on our prejudices, our simple and most basic emotions, and our subliminal propensities. We allow ourselves to be taken in by charlatans of all kinds. We prefer a product that is 80 percent lean to one that is 20 percent fat, and an unnecessary item that costs $9.99 seems much cheaper than one that costs $10. If we hear on the news that the shops will be closed for two days, we rush out to fill our shopping carts as if we were preparing for the famine of the century. We buy everything for twice the price because we have to follow the accepted ritual of Christmas gifts, knowing full well that next week everything will be on sale for half the price. We are willing to get into our cars, stand in lines for two hours, and squash into horrendous shopping centers to save a pittance on a special offer for snacks dripping with sugar and fat. But if we notice that our purchasing power of useless things in that same shopping center starts to drop slightly, we immediately set off for Paris in protest and smash trash cans.

All of these behaviors illustrate our evolutionary inertia. We are descended from animals that had to make fast decisions regarding reproducing at the right time, grabbing an uncertain but valuable food source, escaping a predator by anticipating its moves, and deciding in a matter of seconds whether the person standing in front of you is a friend or enemy. If you want to survive in such an environment, you have to learn to read the signs and clues around you quickly. There is no time to ponder or make improvised probability calculations. Philosophy or science are pretty useless in certain cases. It's best to remember situations that foretell danger immediately and run away. If it is speed that counts, you can afford to be inaccurate. You may get it wrong sometimes, but it is better to have been overly cautious than dead. This is yet another compromise between speed and reliability that generates a cascade of imperfections and split-second decisions.

What philosopher Daniel Dennett, somehow a contemporary heir of Pangloss, has been saying for decades is therefore not true, namely that evolution by natural selection has "designed" us to be rational and almost perfectly logical. Dennett is describing a film, not experimental reality. It *is* true that some particularly gifted human beings have been able to produce refined logical theorems and to conceive of theories of probability and decision-making, but these are not cognitive achievements that come to us spontaneously, "through natural selection." On the contrary, it is precisely cognitive errors, biases, contextual conditioning, and our systematic inability to respect standards of rationality that have an evolutionary root. These flaws derive from the ancient and imperfect compromise between the correctness of our judgments and their speed. Thus irrationality, or at least a limited and pragmatic rationality, has made it possible for us to survive (which does not implicitly mean that it is justified today).

For most of us, meditating at length on the most appropriate action to take in light of the available data, as philosophers or scientists do, is not intuitive. As psychologist Gerd Gigerenzer teaches, from the evolutionary past we have inherited an adaptive, contextual, and ecological rationality, which is not the number one in logic and probabilistic calculation but instead has worked in the real situations in which we had to survive and reproduce. Today we know which adaptive problems, posed by the natural and social environment, this weak and imperfect rationality had to face. It is as if we have an atavistic sensor in our heads that is permanently switched on, and can detect the presence of our fellow beings and predict the moves of potential enemies. For the same reasons, we tend to attribute cause-and-effect relationships between totally unrelated phenomena, such as stepping under a ladder and failing an exam. And this leads us to the endless list of human superstitions.

We are inclined to think that a correlation implies causation. We are prone to make generalizations from anecdotes, or an inadequate number of cases and examples. A great deal of the data from developmental psychology, anthropology, and neuroscience confirms that, for adaptive reasons that no longer exist, our minds have evolved a strong tendency to distinguish between inert entities, such as physical objects, and entities of a psychological nature, like animate agents. We thus are dualists and animists by nature. We then tend to attribute purposes and intentions to both inanimate and animate objects, even when there are none. And again, we connect intentions and purposes, imagining all sorts of plots and conspiracies. For us, stories always have a purpose, which can be evident or hidden.

We are, in other words, belief machines, and we manufacture a lot of those beliefs. And it is not a question of ignorance or dull doctrinal confidence; if we strongly want to believe in something, the more we know, the more we gather information, the more our mind tries to fit the pieces of the puzzle together in order to confirm our original perception. If a bolt of lightning strikes near us, before asking ourselves what the physical cause of the phenomenon was, we presume that someone has sent us a message, and start speculating about who it might be and what the message was. When an improbable event takes place right before our eyes, we immediately think, this cannot be a coincidence! It was destiny. But it was, in fact, a coincidence. Yet if we find belief comforts us, soothes our grievances and bitterness, and gives meaning to this imperfect and senseless story, it becomes indispensable and transforms us into brave champions of delusional theories. We're even willing to pay the price of ridicule, as in the case of the flat-earthers who set out on a cruise to reach the ends of the earth. They will never get there, but afterward, they will come up with a perfect justification to explain why they didn't.

WE ARE NOT EQUIPPED WITH FORESIGHT

There is a simple clue that clearly demonstrates that all of this is an imperfection in the conflicting and shoddy functioning of our brains: we are making a lot of mistakes and even as we do so we are well aware that we are making them. We know perfectly well that we are wrong—or at least, we have all the intellectual and factual tools to understand—but we do it anyway because we like being wrong. But the point is not just that we process thoughts within an archaic emotional soup that often takes control, as if, since the dawn of time, there has been some form of ongoing primordial struggle between reason and instinct. In fact, this is only part of the story.

There is a deeper evolutionary logic that explains (but does not justify) the countless manifestations of human irrationality. Insincerity, the narcissism of gurus, and obscurantism are not enough. Those are only aggravating factors. At the core, there is something more systematic in this paradoxical defect that has helped us survive: it is an imperfection of origin without which we would not be here now talking about it. A large number of scientific studies show that our brain processes its thoughts and decisions using two distinct but interconnected systems (or rather two sets of variously intertwined systems).

Put simply, the first system is old in evolutionarily terms, and presides over our reflexive, everyday, quick, less conscious and pondered actions. We could call it our system of routine, but it functions in emergencies too. It presides over the normal activities that we carry out without noticing day-to-day, from the moment we wake up in the morning, but it also intervenes in dangerous situations in which we need to react quickly without thinking about it for long. In the brain, it is preferably connected to the amygdala, cerebellum, and basal ganglia.

The second system is a recent evolutionarily development and presides over our most deliberate actions, which are the result of a careful and slow evaluation of contextual information. We could call it the logical reasoning system. It presides over activities that require the careful analysis of concepts, generalizations, principles, and abstractions. In the brain, it is preferably connected to the prefrontal cortex. In chapter 5, we saw a conflict between these two systems when the brain is exposed to the vision of a foreign and physically different person.

Neither system is necessarily more rational or emotional than the other. Both can play their roles well, and both have played a fundamental role in our evolution: first, by providing us with instantaneous evaluations based on experience (pointing us in the right direction when coming across a sudden obstacle), which is preferable when the decision has to be taken immediately or is based on a large number of different variables; and second, by offering us the wonders of science and any choice based on reasoned arguments, especially when we are faced with a new problem. Therefore we should not consider one to be irrational and the other rational because reacting instinctively in certain situations is often the most rational choice. At the same time, however, both can cause us to make enormous mistakes because our actions are frequently generated from an improvised middle ground between the two systems.

In fact, neither has definitive control over the other, and the mutual interferences between intuitions and reflections are about as imperfect as it is possible to imagine for a self-proclaimed "*sapiens*" mammal. One tries to control the other, while the other tries to condition it. The deliberative system is, in any case, based on data provided by the reflex system, which is not always reliable. When we are greatly fatigued, or overloaded with work and stress, we need to count on its reliability, but instead it gets bogged down in thousands of cumbersome procedures and misfires, and we can be easily manipulated by those who can cleverly exploit the weaknesses of the reflex system. In other words, when we should use our head, we often react instinctively, and vice versa. When the instinctive and unconscious animallike reaction would be of help, we waste time on trivialities—another evolutionary disparity that generates a cascade of defects.

So we must look for perfection elsewhere, perhaps in our "superior" human faculties. In the last chapter, we saw that it is useless to search for such perfection in our language. What about our memory? Ours is short, tendentiously selective, distorted, and fragmentary. It is nothing more than an animal memory. True, it is quickly activated, and it is excellent at picking out clues from a context, but naturally it must pay the price of unreliability. We don't know exactly where our memories pile up, perhaps all over the brain, but one thing is certain: when they emerge into consciousness, we find they have faded, and we reinterpret them, reconstruct them, amend them, mix them up, or simply lose them. We remember too much or too little. If someone lets us down, we immediately remember all of their past shortcomings, and vice versa. Hence, on the web, we soon forget the source of our information, and we just need to read the same news two or three times to take it for granted once and for all. Our short memories mean that we rarely go back and check how a certain news story turned out. And of course, it means that, historically and individually, we cyclically repeat the same mistakes and horrors.

This is why the decisions we are making today, the probable effects of which will be enjoyed by future generations, are unfortunately not a gift left to us by nature. We have to learn to make decisions through education and culture. In evolution, what counts is here and now. When life expectancy and the quality of life were low, and the difference between life and death lay in grabbing the opportunities of the moment, the future was nothing more than an abstract hypothesis. Consequently, we are not equipped with foresight, and many of us are well aware of this when we arrive home in the evening, see what is on the table, and decide to put off the start of our diets. Numerous experiments have demonstrated that the human mind prefers everything, and now, even if it means getting less in the long run, rather than having to wait for a greater reward in the future (even the near future). It is not rational, but that is how we are. The temptations of the present are irresistible, and feed our deep-seated greed and induced needs. We are more inclined to buy the same object, with the same value, if it is offered to us at a higher price because it seems to be worth more—and yet we called ourselves *Homo economicus*.

And so back to the question, If you were the owner of a company, would you put your trust in such an incoherent and unreliable CEO? Would you put your assets in the hands of a being whose brain, not to mention memory and language, is so poorly constructed that it could go off the rails at any moment? There is obviously no answer because it is like asking the question while facing a mirror. It is a circular paradox. The imperfect *Homo sapiens* has prodigious faculties in calculation, the understanding of physical laws, technological ingenuity, the thirst for knowledge, insatiable curiosity, and the power to transform the environment to suit its own tastes. But the imperfect *Homo sapiens* also has equally prodigious limits when it comes to the body, reasoning, social relations, and limited foresight.

At the bottom of it all, that flaw remains, that imperfection that Holocaust survivor and writer Primo Levi (1971) saw nesting in human nature. People are not beasts, writes Levi (1986) in

The Drowned and the Saved; they become so in certain conditions and contexts that reduce them to following their basic instincts. We inherit an ambivalent and dual nature, so culture and experience can guide us for better or worse. Hence the need for constant ethical and civil vigilance. According to Levi (1985), technical invention, like narrative invention, is tinkering, a composition starting from existing material and its constraints, just like evolution. In this sense, *Homo sapiens* is something totally new for Levi, capable of the sublime but also unthinkable horror. In the appendix to *If This Is a Man*, Levi (1958) writes that the extermination camps are nonhuman, even counter-human inventions, and unprecedented historically. But there can be no return to Arcadia; human beings must continue working as their own blacksmiths. There is only one antidote that can help us avoid reconstructing the infernal termite mounds and falling back into "inhumanism" according to Levi: critical and self-critical rationalism. Not a perfect rationality, but a skeptical and methodical rationalism, whose first lesson is simple: distrust all the prophets that manipulate the imperfections of the human mind.

IN CONCLUSION, THE LAWS OF IMPERFECTION

In the story "The Creation" by Italian writer and journalist Dino Buzzati (2016), after giving birth to the universe, a multitude of the great designer's engineering angels persuade him to also create the earth, a special planet that can host all kinds of plants and animals. In the long process of evaluating the designs of all possible and admirable living oddities, a know-it-all spirit attracts the attention of the Almighty. He has a compelling idea. He shows him the silhouette of a creature that is "rough, clumsy and somewhat indecisive, almost as if the designer, at the crucial moment, had felt discouraged and tired" (all quotes from this source translated by Michael Gerard Kenyon). Although awkward, this being will be the only one gifted with reason—the foolhardy designer assures us—and the only one who can consciously worship his God. The creator, however, wants nothing to do with any know-it-all or intellectual on earth, and dismisses the angel. The latter insists, however, and in the evening, when all the work is done, he takes advantage of the supreme being's fatigue and approaches him, trying again and finally managing to convince him. Even the gods are tempted by curiosity at times. "Also, at the time of creation, optimism was justified," and so the creator signed up for the "fatal project."

We are the children of an absentminded creator named evolution, and on this journey into imperfection, we end up unable to remember anything perfect that came before us. From imperfection to imperfection, from deviation to deviation, here we are, the all-powerful and reckless *Homo sapiens*. Yet if we go back to the beginning, to the beginning of all beginnings, we have come across something perfect: the void. That quantum vacuum full of everything and the opposite of everything, complete and perfect in itself, but timeless. And so that primordial asymmetry, that anomaly, the mother of all imperfections, was necessary to set in motion the history of the universe. It is a history that not only did not foresee us, but that even today appears completely indifferent to our destiny. Should this be true, then perhaps there is a link between time and imperfection.

Darwin understood this: where we find perfection, there is no history. Naturalists wanting to understand how evolution works must look for imperfections, for useless and vestigial traits, because these represent a trace of past changes and the promise of future ones (Gould 1985, 1989). Where we find imperfection, we find something happening—an event, process, change, or relationship. On the contrary, perfection is, by definition, timeless perfection. Where we find perfection, everything has already happened. The gears have no backlash. There are no more alternatives. There is nothing left to tell.

So what if it is precisely our intrinsic imperfection that makes us perceive and experience the sense of time? Evolution means change, events, a contingent happening, an ever-branching exploration of possibilities, and not a necessary progression in absolute time, as the persuasive metaphor of progress would have us believe. In the daily vicissitudes of survival, we have all always been caught up in the perpetual rhythmic alternation of day and night, the seasons and phases of the moon. In this way, we are emotionally immersed in a time that we are convinced we can quantify and measure in sequence because the future is a place where a predator might hide in order to eat us.

According to some physicists, our minds' concept of linear, global, and ordered time might be no more than an enormous and prospective illusion, which for evolutionary reasons has made time an intimate part of the chronobiology of every organism and even every cell (Rovelli 2017). Its irreversible pulsing from the past toward the future pervades us as inhabitants of a peripheral planet where there is nothing, but where everything happens. If evolution feeds on and produces imperfections, then by syllogism it is imperfection that gives us our local sense of time, along with that fastidious, irrepressible feeling of transience. Imperfection is the origin of impermanence. Perfection, on the other hand, is history that has disappeared without a trace.

Medieval philosophers had considered it, and Darwin had seen it in the atrophied wings of penguins and extravagances of barnacles. Perfection is paradoxical. In fact, perfect things cannot be further refined, and only a being who is equally perfect, if not more so, could appreciate them. For something to be perfectible—that is, to have further possibilities of development and completion through the introduction of new characteristics—we would have to admit that it is not currently perfect. So, perfection depends on imperfection. And, indeed, we can only sense it indirectly, through imperfection. Some daring philosophers have therefore gone so far as to theorize that real perfection lies in being imperfect. That is, of course, except for the irascible gods often imagined by us humans—gods who nonetheless frequently derogate from their presumed omnipotence and omniscience by behaving in ways that are anything but blameless.

This divinization of imperfection is unconvincing, as is its sweetening through psychology and counseling. Bookshops are full of publications praising imperfection. There is a potentially infinite number of variations on the recurring theme that "nobody is perfect" (but *è bell'a mamma soja*, above all in Italy).[1] We are advised not to look for impossible perfection because performance anxiety will corrode you. Take comfort in this, and transform your imperfections into resources. Imperfection is natural, of course, but this does not mean that being imperfect is positive, right, or in any way less pleasant. There is no denying that imperfection is also suffering, regret, deluded longing, mortality, waste, and vanity.

If you go for a stroll in the city center at eleven o'clock in the morning on any weekday, when the students are at school and the adults are at work, you suddenly find yourself catapulted into a land of the old. Surrounded by elderly people, as you yourself will hopefully become in a while, you cannot help but think about longevity and the slowing down of human aging that is made possible through medical and social progress and that provides us with the gift of more springs and sunsets—and yet that also comes with a whole load of painful imperfections and inevitable debilitations. The protective mechanisms developed by natural selection do not work so well after we have passed the reproductive age. By delaying aging in this way, we are therefore once again challenging evolution.

We thus fall back on the sixth law of imperfection: the decoupling of the slow pace of biology and fast pace of cultural development. We have more days to live and appreciate this beautiful, senseless world, but only by paying the price of a relentless psychophysical decline punctuated by all sorts of ailments. Other intelligences no less curious and ingenious than ours, such as cephalopods, have moved in the opposite direction

---

[1] This is a Neapolitan saying, which means even a cockroach is beautiful in the eyes of its mother.

(Godfrey-Smith 2016). Cuttlefish and octopuses live only two or three years, with rare exceptions, or almost nothing when compared to us. They endure heavy predation, making their tomorrow uncertain, and so this leads them to reproduce only once, after which they decline. All of that brain, half a billion neurons distributed throughout the body, lives a highly condensed and fleeting experience. Their efficient camera-like eyes (way more perfect than ours!) go opaque, their arms fall off, their three hearts slow down, their bodies sadly fall to pieces, and they let themselves get carried away by the current. We can only imagine that their last moment of consciousness is a farewell to the great ocean from which it all began.

Let me, therefore, take my leave by summarizing my six laws of imperfection:

1. *The law of contingency*: in the form of mutations, genetic drift, mass extinctions, and large-scale ecological changes; chance often unpredictably changes the rules of evolution in such a way that a trait that was previously a big advantage and well defined by natural selection becomes a handicap or a dangerous imperfection.
2. *The law of compromise*: imperfection in nature frequently stems from the need to find a compromise between different interests and antagonistic selective drives and pressures.
3. *The law of constraints*: natural selection is not an agent that perfects and optimizes organisms in every part of their being; it cannot do so because it works in changing contexts, and above all is conditioned by historical, physical, structural, and developmental constraints.
4. *The law of reuse*: the reuse of preexisting structures means that suboptimal and recycled, namely imperfect, structures are frequent in nature.

5. *The law of the onion*: excess, if it can be tolerated, is a source of change because evolution involves the transformation of the possible.
6. *The law of the Red Queen*: when the environment runs faster than we can, we find ourselves evolutionarily out of phase, and thus a little unsuitable or imperfect.

On closer inspection, the six laws have a common feature, which can in no way be considered a consolation: imperfection is a source of "evolvability"—that is, the ability to evolve and generate evolutionary innovations. Plasticity, reuse, and by-products allow organisms to follow new and unpredictable paths from time to time. The ubiquity of imperfection, as Darwin well knew, is the main proof of evolution itself. So, if they tell you that science can only explain the *hows* and not the *whys*, don't believe them. Imperfection is the answer to many *whys*. If they tell you there are no alternatives, don't believe them. We have always coped, as a species, because just at the right time (often on the edge of a precipice), we have looked for and found alternatives.

While writing *Candide* in 1758, Voltaire had clear in his mind the absurdity of both nature and human vicissitudes. A devastating earthquake measuring 8.5 on the Richter scale had razed Lisbon to the ground on All Saints' Day three years earlier, while all the churches were packed full. Europe was being torn apart by the senseless military, political, and religious brutality of the Seven Years' War, and the world was being ravaged by the barbarity and hostilities of colonialism. The universe did not seem to be a machine at all, and even if once it had been, the machinist had long since run for the hills. Yet we should not surrender to human and natural disorder, which two centuries later persists in other forms. The imperfection I have described so far is not the sibling of nihilism. It does not mean that everything is the result of mere chance and that our actions are irrelevant. There is a lot that can be done in a world full of imperfection. We could, for example, transform disenchantment into irony, carve out a shelter, and rebel against at least the remediable human imperfections, such as those that, through the greed and vanity of the few, needlessly generate injustice and inequality for the many. But above all, "we must cultivate our garden," as Voltaire ([1759] 1950) has Candide say twice at the end of his work. If by garden, however, we mean the earth, we have done no such thing, ever, and it is time to start.

Montaigne (1958) writes that humans are miserable beings who are not even lords of themselves and become almost comical when they believe themselves to be lords of the universe. And yet humans are not to be despised for this. Developing an awareness of one's own limits is a long process, which never finishes. Immersed in our fears and defenses, we risk fragmenting our ego and losing our sense of direction long before reaching the serene awareness of our irrelevance as individual, temporarily conscious beings in the great history of the universe—and despite everything, before finding any possible relief in this imperfection.

One among all the others. The critical moments or clinamen of evolution teach us that on several occasions, the past has been open to different outcomes. There was no predetermined path. Many possible counterpresents, more or less imperfect than ours, were not realized, but might have been. So there is no reason to presume that the future will be in any way different. In fact, well aware of the imperfections and potential of the human mind, the great philosopher of science Karl Popper was fond of repeating that the future is also open. Science itself, according to biologist Peter Medawar, is the art of the soluble, and as such it is fruitfully unpredictable. It is up to us to try to influence events to bring about a more desirable and humane counterfuture instead of the rather bleak futures that lie ahead if we continue to listen to our inclinations toward irrationality and fanaticism. Let us not, however, be taken in by presumptuous illusions because it will nevertheless be an imperfect future—but imperfect in a different way.

As we read in the inspired closing lines of Darwin's *The Origin of Species*, there is something amazing in evolution, in this wonderful adventure of life, which in 3.5 billion years has taken us from an amoeba to Donald Trump.

REFERENCES AND FURTHER READING

Ataman, Bulent, Gabriella L. Boulting, David Harmin, Marty G. Yang, Mollie Baker-Salisbury, Ee-Lynn Yap, Athar N. Malik, et al. 2016. "Evolution of Osteocrin as an Activity-Regulated Factor in the Primate Brain." *Nature* 539 (7628): 242–247.

Baggott, Jim. 2015. *Origins: The Scientific Story of Creation*. Oxford: Oxford University Press.

Bowles, Samuel. 2008. "Being Human: Conflict: Altruism's Midwife." *Nature* 456 (7220): 326–327.

Brian Arthur, William. 2009. *The Nature of Technology*. New York: Free Press.

Buzzati, Dino. 2016. "La creazione." In *Il colombre*, 9–14. Milan: Mondadori Editore.

Cavalli-Sforza, Luigi Luca. 2000. *Genes, Peoples, and Languages*. Berkeley: University of California Press.

Cavalli-Sforza, Luigi Luca, and Francesco Cavalli-Sforza. 1995. *The Great Human Diasporas*. Reading, MA: Addison-Wesley.

Cavalli-Sforza, Luigi Luca, and Telmo Pievani. 2012. Homo sapiens*: The Great History of Human Diversity*. Edited by Ian Tattersall. Turin: Codice Editions.

Cipolla, Carlo M. 1988. *The Basic Laws of Human Stupidity*. London: Ebury Publishing.

Coyne, Jerry A. 2009. *Why Evolution Is True*. New York: Penguin Books.

Darwin, Charles R. (1862) 2011. *On the Various Contrivances by Which British and Foreign Orchids Are Fertilised by Insects, and on the Good Effects of Intercrossing*. Cambridge: Cambridge University Press.

Darwin, Charles R. (1871) 1981. *The Descent of Man, and Selection in Relation to Sex*. Princeton, NJ: Princeton University Press.

Darwin, Charles R. 1872. *The Origin of Species by Means of Natural Selection*. 6th ed. London: John Murray.

Darwin, Charles R. (1876) 2009. *The Origin of Species by Means of Natural Selection*. 6th ed. Cambridge: Cambridge University.

Dawkins, Richard. 1986. *The Blind Watchmaker: Why the Evidence of Evolution Reveals a Universe without Design*. New York: W. W. Norton and Company.

Dawkins, Richard. 2004. *The Ancestor's Tale: A Pilgrimage to the Dawn of Evolution*. Boston: Mariner Books.

Dennett, Daniel. 1995. *Darwin's Dangerous Idea: Evolution and the Meanings of Life*. New York: Simon and Schuster.

Dennett, Daniel. 2017. *From Bacteria to Bach and Back: The Evolution of Minds*. New York: W. W. Norton and Company.

DeSalle, Rob, and Ian Tattersall. 2014. *The Brain: Big Bangs, Behaviors, and Beliefs*. New Haven, CT: Yale University Press.

Diamond, Jared. 2005. *Collapse: How Societies Choose to Fail or Succeed*. New York: Viking Press.

Dominguez Bello, Maria G., Rob Knight, Jack A. Gilbert, and Martin J. Blaser. 2018. "Preserving Microbial Diversity." *Science* 362 (6410): 33–34.

Eldredge, Niles. 1995. *Reinventing Darwin: The Great Debate at the High Table of Evolutionary Theory*. New York: John Wiley and Sons.

Eldredge, Niles. 1999. *The Pattern of Evolution*. New York: W. H. Freeman and Company.

ENCODE Project Consortium. 2012. "An Integrated Encyclopedia of DNA Elements in the Human Genome." *Nature* 489:57–74.

Falkowski, Paul G. 2015. *Life's Engines: How Microbes Made Earth Habitable*. Princeton, NJ: Princeton University Press.

Gee, Henry. 2013. *The Accidental Species*. Chicago: University of Chicago Press.

Girotto, Vittorio, Telmo Pievani, and Giorgio Vallortigara. 2014. "Supernatural Beliefs: Adaptations for Social Life or By-products of Cognitive Adaptations?" *Behaviour* 151:385–402.

Godfrey-Smith, Peter. 2009. *Darwinian Populations and Natural Selection*. Oxford: Oxford University Press.

Godfrey-Smith, Peter. 2014. *Philosophy of Biology*. Princeton, NJ: Princeton University Press.

Godfrey-Smith, Peter. 2016. *Other Minds: The Octopus, the Sea, and the Deep Origins of Consciousness*. New York: HarperCollins.

Gould, Stephen J. 1980. *The Panda's Thumb: More Reflections on Natural History*. New York: W. W. Norton and Company.

Gould, Stephen J. 1985. *The Flamingo's Smile: Reflections in Natural History*. New York: W. W. Norton and Company.

Gould, Stephen J. 1989. *Wonderful Life: The Burgess Shale and the Nature of History*. New York: W. W. Norton and Company.

Gould, Stephen J. 1993. *Eight Little Piggies: Reflections in Natural History*. New York: W. W. Norton and Company.

Gould, Stephen J. 2002. *The Structure of Evolutionary Theory*. Cambridge, MA: Belknap Press.

Gould, Stephen J., and Richard C. Lewontin. 1979. "The Spandrels of San Marco and the Panglossian Paradigm: A Critique of the Adaptationist Programme." *Proceedings of the Royal Society of London B: Biological Sciences* 205:581–598.

Gould, Stephen J., and Elisabeth S. Vrba. 1982. "Exaptation—a Missing Term in the Science of Form." *Paleobiology* 8 (1): 4–15.

Graur, Dan, Yichen Zheng, Nicholas Price, Ricardo B. R. Azevedo, Rebecca A. Zufall, and Erfan Elhaik. 2013. "On the Immortality of Television Sets: 'Function' in the Human Genome according to the

Evolution-Free Gospel of ENCODE." *Genome Biology and Evolution* 5 (3): 578–590.

Gregory, T. Ryan, Tyler A. Elliott, and Stefan Linquist. 2016. "Why Genomics Needs Multilevel Evolutionary Theory." In *Evolutionary Theory: A Hierarchical Perspective*, edited by Niles Eldredge, Telmo Pievani, Emanuele Serrelli, and Ilya Tëmkin, 137–150. Chicago: University of Chicago Press.

Henderson, Caspar. 2012. *The Book of Barely Imagined Beings: A 21st Century Bestiary*. Chicago: University of Chicago Press.

Ingraham, John L. 2012. *March of the Microbes: Sighting the Unseen*. Cambridge, MA: Belknap Press.

Jacob, François. 1977. "Evolution and Tinkering." *Science* 196 (4295): 1161–1166.

Jacob, François. 1999. *Of Flies, Mice, and Men*. Cambridge, MA: Harvard University Press.

Kahneman, Daniel. 2011. *Thinking, Fast and Slow*. New York: Farrar, Straus and Giroux.

Kampourakis, Kostas. 2018. *Turning Points: How Critical Events Have Driven Human Evolution, Life, and Development*. New York: Prometheus Books.

Kelly, Kevin. 2010. *What Technology Wants*. New York: Penguin Books.

Kitcher, Philip. 2011. *The Ethical Project*. Cambridge, MA: Harvard University Press.

Kolbert, Elizabeth. 2014. *The Sixth Extinction: An Unnatural History*. New York: Henry Holt and Company.

Kubota, Jennifer T., Mahzarin R. Banaji, and Elizabeth A. Phelps. 2012. "The Neuroscience of Race." *Nature Neuroscience* 15 (7): 940–948.

Landau, Misia. 1991. *Narratives of Human Evolution*. New Haven, CT: Yale University Press.

Leroi-Gourhan, André. 1964. *Gesture and Speech*. Cambridge, MA: MIT Press, 1993.

Levi, Primo. 1958. *Se questo è un uomo*. Turin: Einaudi. Translated by Stuart Woolf as *If This Is a Man* (London: Orion Press, 1959).

Levi, Primo. 1971. *Vizio di forma*. Turin: Einaudi. Translated by Raymond Rosenthal as *The Sixth Day and Other Tales* (New York: Summit Books, 1990); translated by Ann Goldstein and Alessandra Bastagli as *A Tranquil Star* (London: Penguin, 2007).

Levi, Primo. 1985. *L'altrui mestiere*. Turin: Einaudi. Translated by Raymond Rosenthal as *Other People's Trades* (New York: Summit Books, 1989).

Levi, Primo. 1986. *I sommersi e i salvati*. Turin: Einaudi. Translated by Raymond Rosenthal as *The Drowned and the Saved* (New York: Summit Books, 1988).

Levi-Montalcini, Rita. 1987. *In Praise of Imperfection: My Life and Work*. New York: Basic Books.

Lewis, Roy. 1960. *The Evolution of Man: Or How I Ate My Father*. New York: Random House.

Lieberman, Matthew D. 2013. *Social: Why Our Brains Are Wired to Connect*. New York: Crown.

Lucretius. 2008. *De Rerum Natura: The Latin Text of Lucretius*. Edited by William Ellery Leonard and Stanley Barney Smith. Madison: University of Wisconsin Press.

Marcus, Gary. 2008. *Kluge: The Haphazard Evolution of the Human Mind*. Boston: Mariner Books.

Mayr, Ernst. 2001. *What Evolution Is*. New York: Basic Books.

Mayr, Ernst. 2004. *What Makes Biology Unique? Considerations on the Autonomy of a Scientific Discipline*. Cambridge: Cambridge University Press.

Monod, Jacques. 1970. *Chance and Necessity: An Essay on the Natural Philosophy of Modern Biology*. New York: Alfred A. Knopf.

Montaigne, Michel de. 1958. *Complete Essays of Montaigne*. Stanford, CA: Stanford University Press.

Okasha, Samir. 2006. *Evolution and the Levels of Selection*. Oxford: Clarendon Press.

Parravicini, Andrea, and Telmo Pievani. 2016. "Multi-Level Human Evolution: Macroevolutionary Patterns in Hominin Phylogeny." *Journal of Anthropological Sciences* 94:167–192.

Pievani, Telmo. 2011. "Born to Cooperate? Altruism as Exaptation, and the Evolution of Human Sociality." In *Origins of Cooperation and Altruism*, edited by Robert W. Sussman and C. Robert Cloninger, 41–61. New York: Springer.

Pievani, Telmo. 2012a. "An Evolving Research Programme: The Structure of Evolutionary Theory from a Lakatosian Perspective." In *The Theory of Evolution and Its Impact*, edited by Aldo Fasolo, 211–228. New York: Springer.

Pievani, Telmo. 2012b. "Many Ways of Being Human: Stephen J. Gould's Legacy to Paleo-Anthropology (2002–2012)." *Journal of Anthropological Sciences* 90:133–149.

Pievani, Telmo. 2016. "How to Rethink Evolutionary Theory: A Plurality of Evolutionary Patterns." *Evolutionary Biology* 43:446–455.

Pievani, Telmo, and Filippo Sanguettoli. 2020. "The Evolution of Exaptation, and How Exaptation Survived Dennett's Criticism." In *Understanding Innovation Through Exaptation*, edited by Caterina A. M. La Porta, Stefano Zapperi, and Luciano Pilotti, 1–24. Switzerland: Springer Nature.

Pievani, Telmo, and Emanuele Serrelli. 2011. "Exaptation in Human Evolution: How to Test Adaptive vs Exaptive Evolutionary Hypotheses." *Journal of Anthropological Sciences* 89:9–23.

Pigliucci, Massimo, and Gerd B. Müller, eds. *Evolution, the Extended Synthesis*. Cambridge, MA: MIT Press.

Rovelli, Carlo. 2017. *The Order of Time*. New York: Penguin Books.

Sober, Elliott. 1993. *Philosophy of Biology*. Boulder, CO: Westview Press.

Sober, Elliott. 2011. *Did Darwin Write the* Origin *Backwards? Philosophical Essays on Darwin's Theory*. New York: Prometheus Books.

Sterelny, Kim, and Paul E. Griffiths. 1999. *Sex and Death: An Introduction to Philosophy of Biology*. Chicago: University of Chicago Press.

Tattersall, Ian. 1999. *Becoming Human: Evolution and Human Uniqueness*. Boston: Mariner Books.

Tattersall, Ian. 2009. *The Fossil Trail: How We Know What We Think We Know about Human Evolution*. Oxford: Oxford University Press.

Tattersall, Ian. 2012. *Masters of the Planet: The Search for Our Human Origins*. New York: Griffin.

Voltaire. (1759) 1950. *Candide, or Optimism*. Translated by John Butt. London: Penguin Books.

Vrba, Elisabeth S., and Stephen J. Gould. 1986. "The Hierarchical Expansion of Sorting and Selection: Sorting and Selection Cannot Be Equated." *Paleobiology* 12:217–228.

Williams, Robyn. 2006. *Unintelligent Design: Why God Isn't as Smart as She Thinks She Is*. Crows Nest, Australia: Allen and Unwin.

Wilson, Edward O. 2012. *The Social Conquest of Earth*. New York: Liveright.

Wilson, Edward O. 2016. *Half-Earth: Our Planet's Fight for Life*. New York: Liveright.

Zeng, Tian Chen, Alan J. Aw, and Marcus W. Feldman. 2018. "Cultural Hitchhiking and Competition between Patrilineal Kin Groups Explain the Post-Neolithic Y-Chromosome Bottleneck." *Nature Communication* 9 (1): 1–12.

INDEX